AF360918

Finite Element Methods in Smart Materials and Polymers

Finite Element Methods in Smart Materials and Polymers

Editors

Akif Kaynak
Ali Zolfagharian
Saeid Nahavandi

MDPI • Basel • Beijing • Wuhan • Barcelona • Belgrade • Manchester • Tokyo • Cluj • Tianjin

Editors
Akif Kaynak
Deakin University
Australia

Ali Zolfagharian
Deakin University
Australia

Saeid Nahavandi
Deakin University
Australia

Editorial Office
MDPI
St. Alban-Anlage 66
4052 Basel, Switzerland

This is a reprint of articles from the Special Issue published online in the open access journal *Polymers* (ISSN 2073-4360) (available at: https://www.mdpi.com/si/polymers/finite_element_methods).

For citation purposes, cite each article independently as indicated on the article page online and as indicated below:

LastName, A.A.; LastName, B.B.; LastName, C.C. Article Title. *Journal Name* **Year**, *Article Number*, Page Range.

ISBN 978-3-03936-585-2 (Hbk)
ISBN 978-3-03936-586-9 (PDF)

Cover image courtesy of Akif Kaynak.

Contents

About the Editors

Akif Kaynak received his B.Sc. degree from the University of Manchester in UK, M.Sc. degree from Rutgers State University of New Jersey, USA and Ph.D. degree from the University of Technology, Sydney (UTS) in Australia. He is a leading researcher in stimuli-responsive polymers with soft actuators application within the School of Engineering, Deakin University, Australia. He has more than 130 publications, a book chapter, a book on conducting polymers and citations exceeding 2900 and an H-index of 29. He is a regular reviewer for various international journals and part of the advisory board of Sensors journal. He guest edited issues on stimuli responsive polymers in Materials and finite element methods in smart materials and systems in Polymers, MDPI.

Ali Zolfagharian is an Alfred Deakin Medalist for Best Doctoral Thesis, an Alfred Deakin Postdoctoral Fellowship Awardee, at Deakin University, Australia. He is a Mechanical Engineer lecturer with 3D/4D printing of soft robots and soft actuators expertise within the School of Engineering, Deakin University, Australia. He has thus far received \$206k funding to equip his lab from 3DEC (Deakin Digital Design and Engineering Centre), IISRI (Institute for Intelligent Systems Research and Innovation), and an industrial firm combined. His recent research outputs in the field of 3D and 4D printing include guest editing of four special issues in Polymers, Materials, Applied Sciences, one edited book, and 45 articles.

Saeid Nahavandi earned his Ph.D. degree from Durham University, United Kingdom, in 1991. He is an Alfred Deakin professor, Pro Vice-Chancellor (Defence), chair of engineering and the founding director of the Institute for Intelligent Systems Research and Innovation at Deakin University, Australia. His research interests include the modeling of complex systems, robotics, and haptics. He has published more than 900 scientific papers in various international journals and conferences. He is a Fellow of IEEE (FIEEE), Engineers Australia (FIEAust), the Institution of Engineering and Technology (FIET). Saeid is a Fellow of the Australian Academy of Technology and Engineering (ATSE), co-Editor-in-Chief of the IEEE Systems Journal, Associate Editor of IEEE Transactions on Cybernetics and Editor-In-Chief: IEEE SMC Magazine.

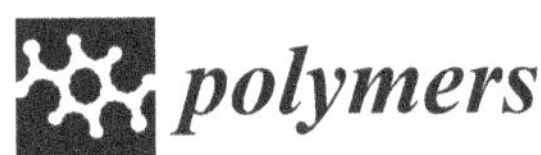 *polymers* 

Editorial

Finite Element Methods in Smart Materials and Polymers

Akif Kaynak [1,*], Ali Zolfagharian [1,*] and Saeid Nahavandi [2]

1   School of Engineering, Deakin University, Geelong, Victoria 3216, Australia
2   Institute for Intelligent Systems Research and Innovation (IISRI), Deakin University, Geelong 3216, Australia; saeid.nahavandi@deakin.edu.au
*   Correspondence: Akif.kaynak@deakin.edu.au (A.K.); a.zolfagharian@deakin.edu.au (A.Z.)

Received: 19 May 2020; Accepted: 23 May 2020; Published: 28 May 2020

Keywords: finite element; smart materials; 3D printing; 4D printing; modelling

---

Functional polymers show unique physical and chemical properties, which can manifest as dynamic responses to external stimuli such as radiation, temperature, chemical reaction, external force, and magnetic and electric fields. Recent advances in the fabrication techniques have enabled different types of polymer systems to be utilized in a wide range of potential applications in smart structures and systems, including structural health monitoring, anti-vibration, and actuators. The progress in these integrated smart structures requires the implementation of finite element modelling using a multiphysics approach in various computational platforms. This special issue presents six scientific report articles.

The Special Issue opens with work from Eindhoven University of Technology in a joint study with Bernal Institute, University of Limerick, Ireland, in which the researchers developed a finite element method (FEM) for representing the microstructural behaviour, particularly the swelling of hydrogel beads [1]. As the design of the cross-linked layers influence the material properties, which, in turn, affect the performance of the hydrogels widely used in pharmaceutical and industrial applications, such as drug delivery or disposable diapers, the authors presented a modelling technique that addressed this limitation for superabsorbent polymers with a partially cross-linked surface layer. The simulations demonstrated that the crack behaviour was influenced by the intrinsic properties of the hydrogel, and provided numerical support for the structural design of the cross-linked hydrogel.

In the second article, Zhang and colleagues from Northeast Agricultural University, China, designed a novel hierarchical metamaterial with a tuneable negative Poisson's ratio by re-entrant representative star-shaped unit cells [2]. In this work, the in-plane mechanical behaviours of the star-re-entrant hierarchical metamaterial were studied, by FEM, in terms of parameters of cell length, angle of inclination, thickness for star subordinate cell, as well as the amount of subordinate cell along x–y directions. The authors claim that the new hierarchical metamaterial will provide further opportunities to design multifunctional lightweight materials that are promising for various engineering applications in the construction, transportation, aerospace, marine, and manufacturing industries due to the inherent low weight associated with hierarchical systems.

Zolfagharian and Kaynak from the School of Engineering in Deakin University, Australia, worked on the fracture resistance analysis of three-dimensional (3D) printed polymers in collaboration with Khoshravani from the University of Siegen in Germany [3]. The researchers investigated the fracture behaviour of 3D-printed plastic components produced by fused deposition modelling (FDM) and multi-jet fusion (MJF) 3D printing techniques to predict the crack propagation leading to catastrophic failure.

U-notched samples manufactured by using nylon and PA12 materials by FDM and MJF printing methods were experimentally analysed. The equivalent material concept (EMC) was used in conjunction with the J-integral failure criterion in ABAQUS software to investigate the failure of the notched samples numerically. Numerical results, supported by experimental analysis, successfully predicted the fracture behaviour of 3D-printed polymer samples. In addition, using the same type of material in the study enabled comparison between the two different printing methods.

In the fourth article, [4], Carleo and colleagues at Queen Mary University of London funded by Jaguar Land Rover investigated the modelling of anti-vibration design in the automotive industry in predicting the dynamic behaviour of the suspension system. In their work, they developed an FEM in ABAQUS software for predicting the viscoelastic behaviour of carbon black-filled rubber as a component of the automotive suspension system. The model used in the study successfully represented the time-dependent phenomenology of filled rubber for use in anti-vibration design considering non-linear elasticity and strain history effects using Maxwell/Prony element and Mullins effect recovery models.

The Special Issue closes with the work of international researchers from the UK, USA, and Egypt in which Atif and colleagues developed a computational fluid dynamics (CFD) model to investigate characteristics of high-speed submicron polyvinylidene fluoride (PVDF) fibres as they are expelled from a blow spinning (SBS) nozzle [5]. The authors used ANSYS Fluent for implementing the CFD model and reported through theoretical and experimental study that a higher air pressure (4 bar) was more suitable to achieve thin fibres of PVDF.

Funding: This research received no external funding.

Acknowledgments: As the Guest Editors, we would like to thank all the authors who submitted papers to this Special Issue. All the papers submitted were peer-reviewed by experts in the field whose comments helped improve the quality of the edition. We would also like to thank the Editorial Board of Polymers for their assistance in managing this Special Issue.

Conflicts of Interest: The authors declare no conflict of interest.

References

1. Ding, J.; Remij, E.W.; Remmers, J.J.; Huyghe, J.M. Effects of intrinsic properties on fracture nucleation and propagation in swelling hydrogels. *Polymers* **2019**, *11*, 926. [CrossRef] [PubMed]
2. Zhang, W.; Zhao, S.; Sun, R.; Scarpa, F.; Wang, J. In-Plane Mechanical Behavior of a New Star-Re-Entrant Hierarchical Metamaterial. *Polymers* **2019**, *11*, 1132. [CrossRef] [PubMed]
3. Zolfagharian, A.; Khosravani, M.R.; Kaynak, A. Fracture Resistance Analysis of 3D-Printed Polymers. *Polymers* **2020**, *12*, 302. [CrossRef]
4. Carleo, F.; Plagge, J.; Whear, R.; Busfield, J.; Klüppel, M. Modeling the Full Time-Dependent Phenomenology of Filled Rubber for Use in Anti-Vibration Design. *Polymers* **2020**, *12*, 841. [CrossRef] [PubMed]
5. Atif, R.; Combrinck, M.; Khaliq, J.; Hassanin, A.H.; Shehata, N.; Elnabawy, E.; Shyha, I. Solution Blow Spinning of High-Performance Submicron Polyvinylidene Fluoride Fibres: Computational Fluid Mechanics Modelling and Experimental Results. *Polymers* **2020**, *12*, 1140. [CrossRef] [PubMed]

Article

Viscoelastic Effects on Drop Deformation Using a Machine Learning-Enhanced, Finite Element Method

Juan Luis Prieto

Escuela Técnica Superior de Ingenieros Industriales, Departamento de Ingeniería Energética, Universidad Politécnica de Madrid, José Gutiérrez Abascal 2, 28006 Madrid, Spain; juanluis.prieto@upm.es

Received: 1 July 2020; Accepted: 21 July 2020; Published: 25 July 2020

Abstract: This paper presents a numerical study of the viscoelastic effects on drop deformation under two configurations of interest: steady shear flow and complex flow under gravitational effects. We use a finite element method along with Brownian dynamics simulation techniques that avoid the use of closed-form, constitutive equations for the "micro-"scale, studying the viscoelastic effects on drop deformation using an interface capturing technique. The method can be enhanced with a variance-reduced approach to the stochastic modeling, along with machine learning techniques to reconstruct the shape of the polymer stress tensor in complex problems where deformations can be dramatic. The results highlight the effects of viscoelasticity on shape, the polymer stress tensor, and flow streamlines under the analyzed configurations.

Keywords: drop; finite element method; machine learning; multiphase flow; particle level set; non-Newtonian fluid

---

1. Introduction

Bubble and drop dynamics in non-Newtonian fluids are a topic of undeniable interest within the community [1], owing largely to the number of real-world situations that may benefit from a comprehensive knowledge of the underlying physics: from drop formation mechanisms [2], to biomedical equipment involving droplet manipulation [3] or engineering devices in which breakup plays a central role [4,5], from droplet impact on liquid surfaces [6] to the study of drop dynamics within polymer gels and solutions [7,8]; a deeper understanding of this type of multiphase flows not only would improve existing manufacturing processes, but also encourage the development of new applications and spur breakthroughs in scientific research.

To study the multiphase flow of polymeric liquids, one should choose an appropriate discretization method capable of providing an accurate description of the interface. Mesh-based methods [9] offer such representation of the interface, either in an explicit (interface-tracking) or in an implicit (interface-capturing) form. Among the former, front-tracking and Arbitrary Lagrangian-Eulerian (ALE) schemes [10,11] display excellent performance in terms of mass conservation and shape preservation; however, remapping techniques are often found necessary [12] under extreme deformation of the moving interface. In contrast, interface-capturing methods follow a Eulerian approach, with the 'Volume-Of-Fluid' (VOF) method positioning itself as one of the most popular techniques within this context [13,14], showing good conservation properties, but requiring additional tools for handling geometrical quantities derived from the interface [15]. As an alternative, Level Set (LS) methods [16,17] capture the interface as the zero isocontour of a certain scalar function, which is advected by the flow, with noticeable mass loss and shape degradation if excessive diffusion is introduced during the advection stage. Despite these shortcomings, the LS method is widely used for interface problems undergoing dramatic deformation and topological changes and can be readily enhanced via "hybrid" schemes such as the Particle Level Set (PLS) method [18] that are able to ease, to a large extent, many of its drawbacks. Whatever the multiphase technique, the correct implementation of appropriate Boundary Conditions (BC) for the configuration of interest is a topic that requires careful consideration of the chosen discretization method [19], as it may have an enormous influence on the actual shape of the interface.

A common approach to handling the polymer interaction with the flow at a macroscopic scale is to represent the polymer contribution to the stress tensor by means of a closed-form, "constitutive" equation; thus, e.g., the work of Yue et al. [20–22] on drop deformation and complex two-phase flow using a diffuse-interface method and constitutive modeling; in [23], Pillapakkam employed an LS method to study rising bubbles in viscoelastic media, while Foteinopoulou and Laso [24] used a Phan–Thien—Tanner model together with an elliptic mesh-deformation algorithm to investigate bubble oscillation; Castillo et al. proposed an LS method with a pressure-enriched FE space to study the two-fluid flow problem along with a Giesekus model for the polymeric liquid [25]; Fraggedakis et al. [26] characterized the critical volume of a bubble rising in a viscoelastic fluid using an FEM-based method and the exponential Phan–Thien and Tanner model; using a coupled LS-VOF ("VOSET") method, Wang et al. [27] studied drag reduction in cavity flow; Xie et al. [28] focused on droplet oscillation under a Maxwell model using lattice Boltzmann techniques. In contrast to constitutive modeling, the "micro-macro" approach [29] tackles the polymer-flow interaction using stochastic and Brownian Dynamics (BD) simulations [30–33] to retrieve the polymer stress tensor from the internal configurations of the polymer particles advected by the flow. Taking the CONNFFESSITapproach of Laso and Öttinger [34], Cormenzana and co-workers [35] and later Grande et al. [36] successfully handled free surface flows of polymer solutions, while Prieto [7,37] conducted multiphase simulations in viscoelastic fluids using a variance-reduced, stochastic implementation of a "micro-macro" method [38]. Further study of multiphase non-Newtonian flows was carried out by Bajaj et al. [39] and Xu et al. [40] using the Brownian Configuration Field (BCF) method of Hulsen et al. [41].

At present, there is a growing interest in Machine Learning (ML) [42,43] in the context of polymer simulation: Doblies et al. [44] employed ML as a means of predicting mechanical properties, while Jackson and collaborators [45] focused on the optoelectronic properties of conjugated polymers. It was also used by Kopal and co-workers in the prediction of the viscoelastic behavior of elastomer systems [46] using Radial Basis Functions (RBFs) or as a data-driven classification method to determine polymer/solvent compatibility by Chandrasekaran et al. [47]. Fluid dynamics of multiphase flow also benefits from this recent focus on ML: Ma and collaborators used statistical learning in [48] for bubbly systems; Ladický et al. accelerated multiphase simulations in GPUs using a data-driven approach with regression forests [49]; and Gibou and co-workers employed deep learning techniques with sharp interface methods in [50]. Hence, it seems unquestionable that ML and data-driven simulations are already having a tremendous impact within the scientific community, and the future possibilities seem nearly endless.

The main purpose of this paper is to highlight the effects caused by the non-Newtonian behavior of the polymer solution in a multiphase flow system, performing a series of numerical simulations by means of a computationally efficient, accurate, and robust numerical method based on a finite element discretization of the governing equations that uses ML-inspired techniques for the reconstruction of the polymer stress tensor, a feature that proves to be of the greatest importance for the accurate characterization of the viscoelastic flow. Thus, after presenting this Introduction, we describe the governing equations along with the aspects of the computational implementation in the Materials and Methods Section 2. Then, we move to Section 3, where we present results for drop deformation under steady, shear flows and in situations where gravitational effects drive the dynamics of the flow. Finally, Section 4 offers some conclusions and future lines of work.

2. Materials and Methods

In this section, we describe the main ideas, mathematical background, and computational implementation of the FEM-based, ML-enhanced method used to perform the series of numerical experiments carried out in Section 3 to gain insight into the impact that viscoelasticity may have on drop deformation. We start with the finite element discretization, then moving on to the computational implementation.

2.1. Finite Element Discretization

The problem of drop deformation in polymer solutions can be tackled using different methods. In this work, we use an FEM-based discretization of the "macro scale" equations governing the fluid flow, employing a stochastic modeling of the stresses arising from the polymer fluid to account for the "micro"-scale. Finally, we use a Particle Level Set (PLS) method to capture the interface of the deforming drop.

2.1.1. Governing Equations

The two-phase, incompressible fluid flow studied in this work is governed, at the "macro"-scale, by the Navier–Stokes equations, which in dimensionless form can be written in the space-time domain $\Omega \times [0, T]$ as:

$$
\begin{cases}
Re\rho \dfrac{D\mathbf{u}}{Dt} - \nabla \cdot \left[\mu \left(\nabla \mathbf{u} + (\nabla \mathbf{u})^T \right) \right] + \nabla p = -\rho \mathbf{e}_z \dfrac{Re}{Fr^2} + \dfrac{c}{De} \nabla \cdot \boldsymbol{\tau}_p + \dfrac{Re}{We} \kappa \delta_\Gamma(\phi) \mathbf{n}, \\[2mm]
\nabla \cdot \mathbf{u} = 0; \\[2mm]
\mathbf{u}(\mathbf{x}, 0) = \mathbf{u}_0(\mathbf{x}) \quad \forall \mathbf{x} \in \Omega, \\[2mm]
\mathbf{u}(\mathbf{x}, t) = 0 \quad \text{on} \quad \delta D_{\text{no-slip}} \subset \delta\Omega, \forall t \in (0, T), \\[2mm]
\mathbf{u}(\mathbf{x}_{\text{img}}, t) = (\mathbf{x}_{\text{ref}}, t) \quad \text{on} \quad \delta D_{\text{PBC}} \subset \delta\Omega, \forall t \in (0, T), \\[2mm]
\mathbf{u}(\mathbf{x}, t) \cdot \mathbf{n} = 0 \quad \text{and} \quad \mathbf{n} \cdot \boldsymbol{\tau}_s \cdot \mathbf{t} = 0 \quad \text{on} \quad \delta D_{\text{free-slip}} = \delta\Omega \setminus (\delta D_{\text{no-slip}} \cup \delta D_{\text{pbc}}), \forall t \in (0, T).
\end{cases}
\tag{1}
$$

where ρ is the density, μ the viscosity, $D()/Dt$ the total (convective) derivative operator, p the pressure, $\mathbf{v}$ the velocity, $\boldsymbol{\tau}_p$ the polymer ("extra-")stress tensor, Γ the interface of the drop, κ its curvature, ϕ the level set function (see Section 2.1.2), Re the Reynolds number, Fr the Froude number, We the Weber number, c the concentration parameter, De the Deborah number, T the final time, and $\delta\Omega$ the boundary of the spatial domain Ω. The "no-slip", "free-slip", and "Periodic Boundary Conditions" (PBC), are implemented in Section 2.2.3 for the configurations explored in the Results and Discussion Section 3. Using a semi-Lagrangian approach, the total convective operator in Equation (1) is discretized along the "characteristic curves" of the flow without changing the Eulerian, underlying mesh. The resulting weak formulation can be written as a Stokes-like problem to be solved at each instant of time, with the following form:

$$
\begin{cases}
\dfrac{3Re}{2dt} \left(\rho^*(\phi_h^n) \mathbf{u}_h^n, \psi_h \right) + \left(\mu^*(\phi_h^n) \nabla \mathbf{u}_h^n, \nabla \psi_h \right) - \left(p_h^n, \nabla \cdot \psi_h \right) = \dfrac{2Re}{dt} \left(\rho^*(\phi_h^n) \bar{\mathbf{u}}_h^{n-1}, \psi_h \right) - \\[3mm]
- \dfrac{Re}{2dt} \left(\rho^*(\phi_h^n) \bar{\mathbf{u}}_h^{n-2}, \psi_h \right) - \dfrac{Re}{Fr^2} \left(\rho^*(\phi_h^n) \mathbf{e}_z, \psi_h \right) + \\[3mm]
+ \dfrac{c}{De} \left(\nabla \cdot \boldsymbol{\tau}_{ph}^n, \psi_h \right) + \dfrac{Re}{We} \left(\kappa_h^n \delta_{\Gamma_{h/2}}(\phi_h^n) \mathbf{n}_h^n, \psi_h \right), \forall \psi_h \in \mathbf{V}_{h0}; \\[3mm]
\left(\nabla \cdot \mathbf{u}_h^n, q_h \right) = 0, \quad \forall q_h \in Q_h; \qquad \text{with} : (a, b) \equiv \displaystyle\int_D ab\, dx
\end{cases}
\tag{2}
$$

where the subscript h denotes the spatial discretization of the variables; the superscripts $*$ and n their non-dimensionality and temporal discretization at instant of time n, respectively; dt is the time step size used in the discretization of the full interval $(0, T)$; ψ_h the basis functions of velocity vector space $\mathbf{V}_h$ and q_h the basis functions of the pressure space Q_h. The basis functions of the finite element spaces are chosen to be polynomials of degree $\{\mathbb{P}_2, \mathbb{P}_1, \mathbb{P}_2, \mathbb{P}_1\}$ for velocity, pressure, the level set function, and the polymer stress tensor, respectively, thus satisfying the Ladyzhenskaya–Babuška–Brezzi condition.

The spatial discretization of the domain Ω is accomplished using an unstructured mesh composed of simplices of average size h, and the surface tension effects are implemented making use of the Laplace–Beltrami operator, as detailed in [37].

In Equation (2), we collect the "macro-scale" physics involved in the FEM-based discretization of a multiphase flow of incompressible fluids subject to gravity, viscous, and surface tension forces, in which the interface is retrieved by a level set function ϕ, and the viscoelastic effects of the polymer solution are captured through the polymer stress tensor τ_p. We now take a stochastic approach to the computation of τ_p avoiding closed-form constitutive equations, performing instead Brownian dynamics simulations of polymer particles [7] modeled via the Hooke (equivalent to the Oldroyd-B constitutive equation) and FENE (Finitely Extensible Non-linear Elastic) kinetic models [51]. To compute τ_p, we need to average over the internal configurations $\mathbf{Q}$ of each of the N_d dumbbells ("polymer particles"), which are scattered over the domain and follow the flow. Thus, we solve for each of the dumbbells the following stochastic (dimensionless) differential equation:

$$
d\mathbf{Q} = \begin{cases} \left(\kappa \cdot \mathbf{Q} - \dfrac{1}{2De}\mathbf{Q} \right) dt + \dfrac{1}{\sqrt{De}}d\mathbf{W}, & \text{Hooke;} \\[4mm] \left(\kappa \cdot \mathbf{Q} - \dfrac{1}{2De}\dfrac{\mathbf{Q}}{1 - \|\mathbf{Q}\|^2/b} \right) dt + \dfrac{1}{\sqrt{De}}d\mathbf{W}, & \text{FENE;} \end{cases} \tag{3}
$$

a task that is accomplished by means of a (weak) second-order accurate predictor-corrector algorithm proposed by Öttinger [51]. In Equation (3), κ is the velocity gradient, $\mathbf{W}$ a stochastic Wiener process, the Deborah number De representing the ratio between the relaxation time of the polymer and a characteristic time of the flow, and b is the FENE extensibility parameter so that, as $b \to \infty$, Hooke→FENE. A variance-reduction technique can also be implemented [37,41], performing the average over the N_d dumbbells with uncorrelated noise that comprise each of the N_{ens} ensembles, so that the i-th dumbbell of each ensemble is subject to the same (correlated) stochastic "kick", reducing the overall noise of the simulation. Finally, the polymer stress tensor is reconstructed and evaluated at the mesh nodes [52] using the ML-inspired techniques of Section 2.2.1.

2.1.2. Interface Capturing Technique

The interface Γ is captured as the zero isocontour of a level set function ϕ, which is transported by the characteristic curves of the flow through the total derivative operator:

$$
\frac{D\phi}{Dt} = \frac{\partial \phi}{\partial t} + \mathbf{v} \cdot \nabla \phi = 0. \tag{4}
$$

As we did with Equation (1), the convective terms in Equation (4) are dealt with using a semi-Lagrangian approach, as proposed first in [53]. The level set function ϕ is initialized as a signed-distance function to improve the regularity and behavior of the transported solution. However, as time goes on, discretization errors build up in the form of numerical artifacts that may eventually propagate into the interface and destroy it. To prevent this undesired effect, an Eikonal-based, redistancing procedure [54] is performed at each time step, so that we ensure that, within a band around the interface, the level set function preserves the signed-distance property. To improve mass conservation and shape definition, we add N_{mp} marker particles [18] that are passively advected by the flow, helping to correct the interface by defining local level set functions with a variable radius $0 \le r_{\min} \le r_p^n \le r_{\max} < h$, accounting for lost resolution in the sub-grid scales (see [37] for details).

2.2. Computational Implementation

Next, we describe some of the topics involved in the computational implementation of the ML-enhanced, FEM-based method utilized to tackle the drop deformation problems proposed in Section 3. All code was written in the C programming language and compiled using the open-source GNU-C compiler gcc. Efficiency and robustness are two features pursued in this implementation;

accordingly, all simulations presented in this work can be run in a commodity personal computer with a 4-core processor. All 2D plots were made using the open-source library `matplotlib` [55] and/or the TikZ and PGF packages [56], while the 3D graphics make use of Mayavi [57]. For data manipulation and data analysis, we use the `pandas` [58] package.

2.2.1. Machine Learning Enhancement

Radial basis functions are a family of kernel methods [59] widely used in Machine Learning (ML) for their flexibility in generating spaces of trial functions with excellent approximation properties. This close relation to approximation theory allows for a solid mathematical background, which proves extremely useful when offering bounds in supervised learning problems [43] for the necessary number of training data required to provide a model trained with such data and having a small generalization error. Combining these concepts with novel applications in the field of image processing [60,61], we leverage RBFs to reconstruct the discrete solution of the polymer stress tensor over the whole domain, using the scattered data provided by the values of the polymer stress tensor defined at each of the position of the ensembles. The main advantage of this idea is the mesh independence property guaranteed by the RBF approach; the relevance of this feature cannot be overstated in the context of "micro-macro" methods for non-Newtonian flows [29,34,41]. Traditional methods use particles for the kinetic modeling of the polymer solution; however, spatial mesh refinement hinders the accuracy of the stochastic model (given the reduced number of particles surrounding a certain mesh node), thus requiring compromising on the accuracy of either the "micro-" (internal configurations of the polymer and "extra-"stress tensor) or the "macro-" (velocity, pressure) scales. This is no longer the case with our approach: now, using the smooth approximation built via RBFs, we can evaluate the tensor at any position (e.g., a certain mesh point), virtually decoupling the "micro-" and "macro-"scales and paving the way for adaptive mesh refinement in "micro-macro" methods.

Thus, at each instant of time, we retrieve the positions $\{x_i\}, i = 1, .., N_{ens}$ of all the ensembles scattered over the domain. Then, we compute $\tau_{p,i}$, the value of the polymer stress tensor at the ensemble i using Kramers' expression averaging over the N_d dumbbells contained at the ensembles:

$$\tau_{p,i} = nk_B\Theta \left[\frac{1}{N_d} \sum_{j=1}^{N_d} \mathbf{F}\left(\mathbf{Q}_j^i\right) \otimes \mathbf{Q}_j^i - \mathbf{I} \right]. \tag{5}$$

As kernels of approximation, we focus on compactly-supported RBFs [59,62] for their computational efficiency and remarkable accuracy, since the matrices resulting from a suitable formulation of the approximation problem are sparse. In particular, we use Wendland's $\varphi_{3,k}(C^{2k}), k = 0, .., 3$, suitable for problems of space dimension $d \leq 3$, and degree of smoothness $2k$:

$$\begin{cases} \varphi_{3,0}(r) = (1-r)_+^2, \\ \varphi_{3,1}(r) = (1-r)_+^4 (4r+1), \\ \varphi_{3,2}(r) = (1-r)_+^6 (35r^2 + 18r + 3), \\ \varphi_{3,3}(r) = (1-r)_+^8 (32r^3 + 25r^2 + 8r + 1), \end{cases} \tag{6}$$

so that outside the support size χ, $\varphi_{3,k}(r > \chi) = 0$, whereas inside the support, they are positive definite on $\mathbb{R}^d$ and provide optimal convergence rate $\mathcal{O}(h_\Omega)^{2k+d+1}$, with h_Ω the fill distance [62]. Using these kernels, we build an interpolant s of the form:

$$s(\mathbf{x}) = p(\mathbf{x}) + \sum_i \lambda_i \varphi(\|\mathbf{x} - \mathbf{x}_i\|), \tag{7}$$

with $p(\mathbf{x})$ a quadratic function defined by coefficients c_i and and λ_i a set of real-valued interpolation coefficients for the CSRBFs. The formulation can then be reduced to a system of equations equivalent to a saddle-point problem of the form `[Q P; Pᵀ 0]] [1; c] = [f; 0]`, which can be solved efficiently using the two-step approach described in [52] to obtain the solution vectors [1;c] required to build the interpolant *s*. After building the interpolant and having reconstructed the polymer stress tensor, we can straightforwardly evaluate each component at any given point (e.g., at one mesh node).

The use of CSRBFs as approximation kernels calls for the determination of the support size χ, for each of the ensembles scattered over the domain. The technique used to perform this potentially expensive computation is taken from the classification problems usually found in ML: the Nearest Neighbor (NN) classification algorithm [42,43]. Using this approach, along with an implementation of kd-tree as data structures, we are able to obtain, at each instant of time, the nearest neighbors of each ensemble, within a certain support size χ.

2.2.2. PETSc-Based Solver

The solution of the linear systems arising from the FEM discretization of the coupled velocity-pressure system is carried out taking advantage of PETSc, the Portable, Extensible Toolkit for Scientific Computation [63]. This step is key to the overall efficiency and viability of the method as an appropriate candidate to tackle Newtonian and non-Newtonian multiphase flows in which large density and viscosity ratios are involved, as those taking place in experiments dealing with drop and bubble deformation. As the ratio increases, so does the condition number of the resulting velocity matrix, which hampers the convergence of traditional iterative approaches to the solution of the coupled velocity-pressure system such as preconditioned conjugate gradient techniques [64] or traditional "splitting" techniques, usually problematic at low (viscosity-dominant) Reynolds numbers [65]. Instead, we use a physics-based, block-preconditioning approach presented by Elman and collaborators [66] and implemented in the PETSc-FieldSplit preconditioner: we define separate fields for the discrete velocity and pressure and combine them using a Schur-complement factorization [67] of the total system matrix, which is efficiently stored as a MatNest structure K=[A B; B^{T} 0], in which the corresponding sub-matrices for the discrete velocity, discrete gradient, and discrete divergence are the four entries K_ij, for i,j=1,2. The following snippet collects a sample function to build the block matrix.

```
PetscErrorCode fun_petsc_construct_block_matrix_PETSC_Stokes_system (...)
{
  ...
  /* Build first sub-matrix "K00" */
  ierr = MatDuplicate(K00,MAT_COPY_VALUES,& (Stokes_Problem_system->K_ij[0]));CHKERRQ(ierr);
  ierr = MatAssemblyBegin(Stokes_Problem_system->K_ij[0],MAT_FINAL_ASSEMBLY); CHKERRQ(ierr);
  ierr = MatAssemblyEnd(Stokes_Problem_system->K_ij[0],MAT_FINAL_ASSEMBLY); CHKERRQ(ierr);
  ...
  /* Build complete, block matrix "K" from sub-matrices "K_ij" as a nested matrix */
  ierr = MatCreateNest(PETSC_COMM_WORLD, 2, NULL, 2, NULL, \
          Stokes_Problem_system->K_ij, & (Stokes_Problem_system->K)); CHKERRQ(ierr);
  ierr = MatAssemblyBegin(Stokes_Problem_system->K,MAT_FINAL_ASSEMBLY); CHKERRQ(ierr);
  ierr = MatAssemblyEnd(Stokes_Problem_system->K,MAT_FINAL_ASSEMBLY); CHKERRQ(ierr);
  ...
}
```

The previous approach can now be used to tackle the saddle-point problem resulting from our FEM-based discretization of the Navier–Stokes equations using the PETSc-FieldSplit preconditioner for the system matrix, collecting all the information in a suitable structure such as the PETSc_Saddle_point_system sample structure included in the following code snippet.

```c
/* Structure with all the information required to solve a saddle-point problem */
typedef struct
{
  Mat            K;         /*MatNest matrix with 4 sub-matrices: K = [K00 K01; K10 K11]*/
  Mat            K_ij[4];        /*each of the four sub-matrices of block matrix "K"*/
  Mat            Pmat;         /*MatNest precond-matrix: Pmat = [Pmat00 Pmat01; Pmat10 Pmat11]; */
  Mat            Pmat_ij[4];       /*each of the four sub-matrices of precond-matrix "Pmat"*/
  Vec            x;        /*solution of system*/
  Vec            b;        /*right hand side vector*/
  KSP            ksp; /*solver context*/
  MatNullSpace nullsp; /*Nullspace for pressure (in Stokes solved by PCFieldSplit)*/
  PC             pc; /*preconditioner context*/
  IS             isg[2]; /* index sets (rows) of splits "0" and "1" (e.g. velocity, pressure)*/
}
PETSC_Saddle_point_system;
```

The CHOLMOD package [68] is used to perform the sparse direct Cholesky factorization of the positive-definite block matrix K00=A, while the Least-Squares Commutator (LSC) preconditioner [67] takes care of the second field (pressure) involved in the upper factorization of the Schur complement, using a combination of the conjugate gradient and additive-Schwarz methods [69] for the multigrid levels of the ML (Multi-Level) preconditioner package [70] used for LSC. The following options passed to PETSc were found to offer excellent performance in demanding multiphase flow problems with high density ratios such as those found in Section 3.2.

```c
/* KSP00=preonly (cholesky,CHOLMOD); KSP11:preonly (lsc); ksp_lsc: cg (ml(+asm)). */
char  options_pc_stokes[] = "-stokes_ksp_rtol 5.e-9 -stokes_ksp_diagonal_scale \
-stokes_ksp_type fgmres -stokes_pc_type fieldsplit -stokes_pc_fieldsplit_type schur \
-stokes_pc_fieldsplit_schur_fact_type upper -stokes_pc_fieldsplit_detect_saddle_point \
-stokes_fieldsplit_0_pc_type cholesky -stokes_fieldsplit_0_pc_factor_mat_solver_package cholmod \
-stokes_fieldsplit_0_ksp_type preonly -stokes_fieldsplit_1_pc_type lsc -stokes_fieldsplit_1_lsc_pc_type ml \
-stokes_fieldsplit_1_lsc_mg_coarse_pc_factor_shift_type NONZERO \
-stokes_fieldsplit_1_lsc_mg_levels_1_pc_type asm -stokes_fieldsplit_1_lsc_mg_levels_2_pc_type asm \
-stokes_fieldsplit_1_lsc_mg_levels_3_pc_type asm -stokes_fieldsplit_1_lsc_mg_levels_4_pc_type asm \
-stokes_fieldsplit_1_lsc_mg_levels_5_pc_type asm -stokes_fieldsplit_1_lsc_ksp_max_it 3 \
-stokes_fieldsplit_1_lsc_ksp_type cg  -stokes_fieldsplit_1_lsc_ksp_constant_null_space  \
-stokes_fieldsplit_1_ksp_type preonly";
```

These are some of the general configuration options and data structures that are most responsible for the performance boost we find for the kind of complex, polymer flows studied in this work. Given the platform-independent model of C and PETSc, the snippets could be included with little modification in any existing C code used by any researcher with a similar underlying mathematical problem, independently of the particular field of knowledge.

2.2.3. Boundary Conditions

The viscous and viscoelastic flows presented here for the analysis of drop deformation in polymer solutions make use of several types of boundary conditions: the "no-slip" boundary condition to ensure that the fluid adheres to the adjacent solid boundary; the "free-slip" boundary condition to neglect friction forces at the fluid-solid interface; and Periodic Boundary Conditions (PBCs) at the inlet and outlet of the domain to model sufficiently large domains.

In our FEM-based method, the "free-slip" condition can be applied naturally as a Neumann-type condition that is satisfied automatically by the weak formulation of the problem [7,37]. However, the implementation of PBCs in unstructured meshes for the configuration presented in Section 3.1 for drop deformation in shear flows is not straightforward. Following Pask et al. [71] and Sukumar and Pask [72], we carry out an efficient implementation of the PBCs in structured and unstructured finite element meshes using the idea of "reference" and "image" nodes at the inlet and outlet, respectively (see Figure 1). First, we obtain the connectivity lists for the assembled sub-matrices K_00,K_01,K_10 of the block matrix K = [K00 K01; K10 K11]; then, we perform row operations (first) and column operations (later) to K00,K01,K10 and to the right-hand side F, as described in [72]; finally, we update the values of the (pressure,velocity) "image" nodes with the values obtained from the corresponding "reference" nodes. The addition of dumbbells and marker particles to deal with multiphase flows and non-Newtonian fluids adds complexity to the actual implementation, but not to the general concept.

As for the "no-slip" conditions, they translate into homogeneous or non-homogeneous Dirichlet-type conditions that can be applied efficiently using the symmetry-conserving procedure `MatZeroRowsColumns` to perform row and column operations on the PETSc-based assembled matrices `K00,K01,K10` and right-hand side `F`.

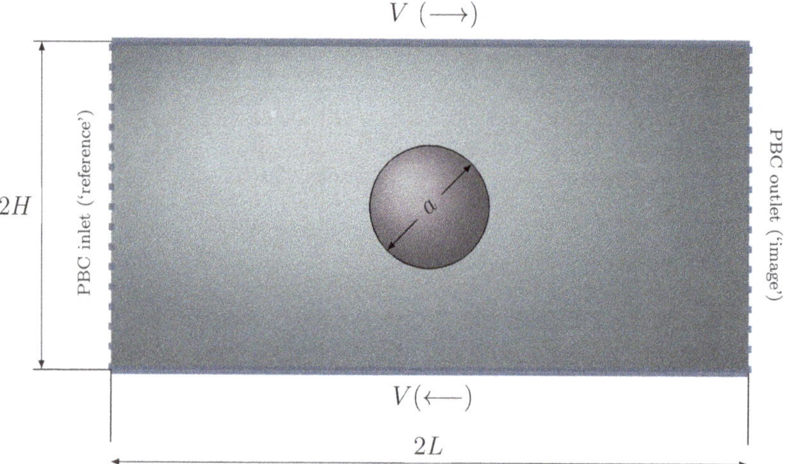

Figure 1. Schematic of the computational domain for a drop deforming in shear flow.

3. Results and Discussion

The main purpose of this paper is to explore the behavior of drops immersed in polymer solutions through a series of numerical experiments that try to highlight the effects of viscoelasticity in the shape of the drop, as well as in the emerging flow patterns. First, we focus on drop deformation under a steady, shear flow, offering results under two configurations: Newtonian drop in a Newtonian "matrix" (ambient fluid) and non-Newtonian drop in a Newtonian matrix. Then, we make use of our ML-enhanced, FEM-based method to investigate stronger viscoelasticity effects in flows where gravitational forces and surface tension effects can become relevant, showing the influence that the discretization employed and the numerical techniques utilized may have on the accurate reconstruction of the polymer stress tensor and on the quality of the discrete solution for the multiphase flow.

3.1. Drop Deformation in Steady, Shear Flow

We start the exploration of viscoelastic effects on drop deformation by considering the problem of a drop immersed in another, ambient fluid and subject to a steady, shear flow. First, we show results for a Newtonian/Newtonian system; then, we introduce viscoelasticity in the system, comparing our results with others obtained using a constitutive (Oldroyd-B) model, equivalent to our stochastic, Hookean dumbbell approach. The configuration is represented in Figure 1: a drop of a viscous (real) fluid of radius a placed at the center of a domain $[2L \times 2H]$, with $H = 4a$ and $L = 8a$, experiences a steady, shear flow of rate $\dot{\gamma} = V/H$ produced by the top and bottom lids moving at velocity V in opposite directions.

The inlet and outlet walls have Periodic Boundary Conditions (PBCs), the implementation of which is detailed in Section 2.2.3. At the top and bottom panels, the "no-slip" boundary condition is considered.

3.1.1. Newtonian Drop in a Newtonian Matrix

We place a Newtonian drop in a Newtonian ambient fluid as depicted in Figure 1 to study the shape of the interface for different values of the capillary number $Ca = We/Re$ in terms of the "deformation parameter" D. This parameter is usually defined as $D = (L - B)/(L + B)$, with L and B being the longest and shortest lengths from the center of the drop to the interface (corresponding also to the major and minor axes of the ellipse), respectively. We also choose a small value for the Reynolds number ($Re = 0.1$) so as to neglect strong inertial effects, comparing our results with those of Zhou and Pozrikidis [73], Yue et al. [20,22], and Afkhami et al. [74]. We employ an unstructured, uniform mesh of size $h = 1/80$, in a rectangular domain $[0,2] \times [0,1]$ with time step $dt = 1/100$ and $N_{mp} = 2.5 \times 10^5$ marker particles. In Figure 2, we represent the steady value of the deformation parameter for increasing values of the capillarity number; in all cases, we obtain excellent agreement with the published results, especially with that by Afkhami et al. [74], deviating in the worst case scenario $< 2.5\%$ at $Ca = 0.6$ from the results offered by Zhou and Pozrikidis in [73]. The steady values of D show a linear behavior with Ca for small capillary numbers, then decreasing as Ca gets larger; as pointed out in [20], we expect the physical insights gained by this 2D exploration to be relevant to actual (three-dimensional) experiments.

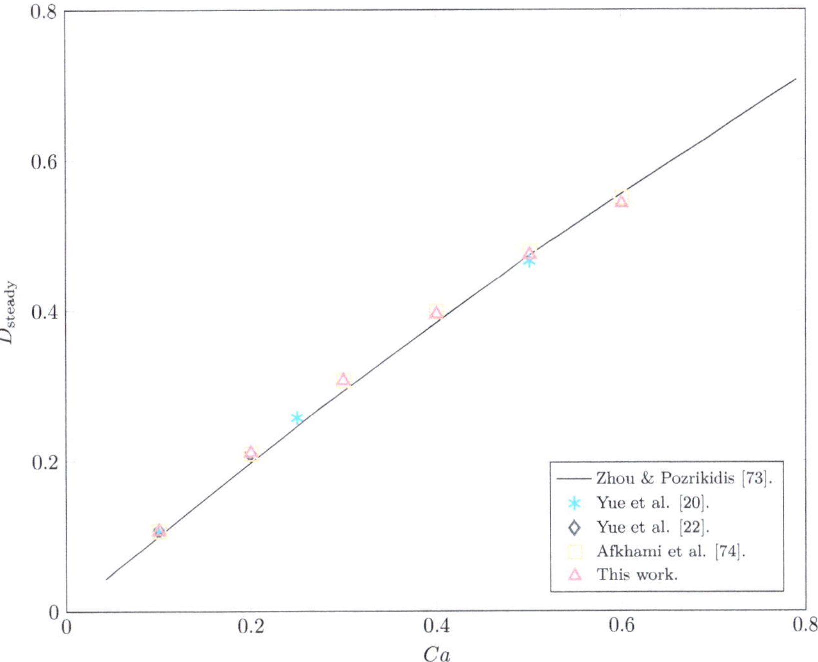

Figure 2. Comparison of the steady-state deformation of a Newtonian drop in a Newtonian matrix between our results and those found in Zhou & Pozrikidis [73], Yue et al. [20,22], and Afkhami et al. [74].

We also collect in Figure 3 the evolution of the deformation parameter D as a function of the dimensionless time $t^* = t\dot{\gamma}$ and compare it with the results of [74], for a wide range of capillary numbers from $Ca = 0.1$ to $Ca = 0.6$. Again, an excellent agreement is obtained throughout, with a maximum difference of less than 1.3% (for $Ca = 0.6$) between the two numerical methods at any given time.

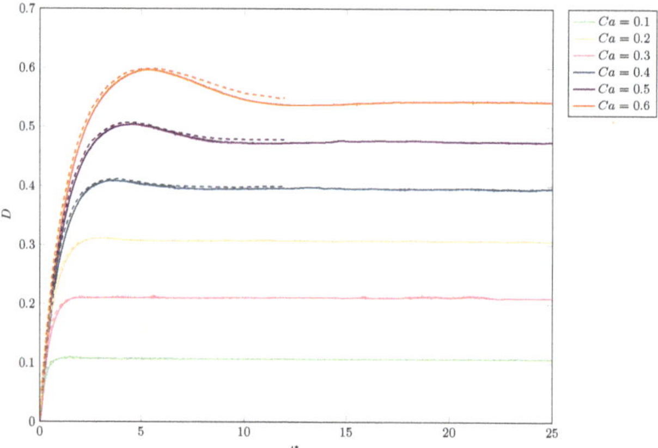

Figure 3. Comparison of the evolution of deformation for a Newtonian drop in a Newtonian matrix of equal viscosity, for different values of capillarity *Ca*, between our results (solid line) and those found in Afkhami et al. [74] (dashed line).

Following Yue et al. [22], we now study the steady shape of the drop in polar coordinates, collecting the radius r for each orientation angle $\theta \in [0, \pi]$, with θ defined as the angle formed by the line connecting the center of gravity of the drop and a point in the free-surface, with the x-axis. Thus, we show in Figure 4 the shape in polar coordinates and the actual shape for a simulation with $Ca = 0.1$.

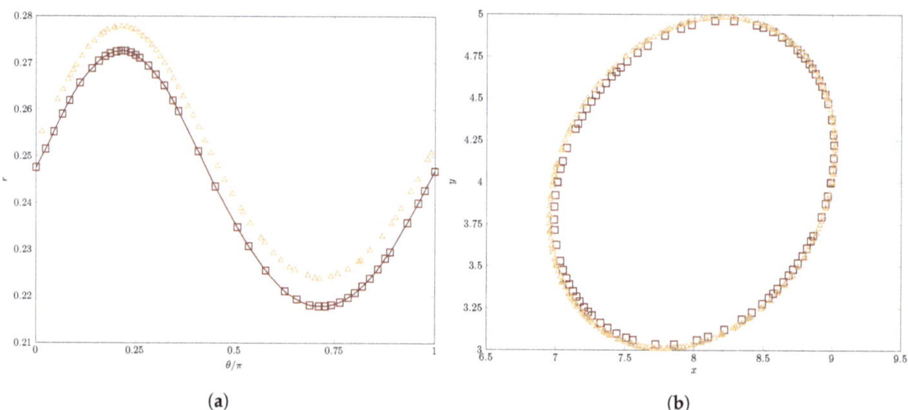

(**a**) (**b**)

Figure 4. Comparison of the drop shape in (**a**) polar coordinates and (**b**) the actual shape, for $Ca = 0.1$, between our results (triangles) and those found in Yue et al. [22] (square markers).

We can observe in this Figure 4 very good agreement in the orientation angle of the droplet, which can be measured at $\hat{\theta} \approx 39.6°$, lower than $45°$ as numerical and experimental results confirm. However, there is a major discrepancy between the results of the two methods: the radii for each polar angle are smaller for the method of Yue et al. [22] when compared with ours (see left panel). To better understand what is happening here, we plot the actual shape of the drop according to both methods in the right panel of Figure 4. As can be observed, the reason for the discrepancy lies in the lack of mass conservation "mass-loss" incurred by the drop of [22], which loses around 5% of its initial mass at the end of the simulation, whereas we can ensure mass conservation up to 99.98% of the initial mass in this simulation.

3.1.2. Viscoelastic Drop in a Newtonian Matrix

After exploring the multiphase Newtonian/Newtonian system and comparing our results with those of the literature, we now proceed with a problem in which the viscoelastic effects are present. We focus on a viscoelastic drop modeled by the Hookean dumbbell model (equivalent to the Oldroyd-B constitutive equation), immersed in a Newtonian, ambient fluid, using the same configuration of the previous Newtonian/Newtonian analysis. The flow is impulsively started at $t = 0$ with shear rate $\dot{\gamma} = 1$, using a rather coarse mesh with 80×40 elements and $N_p = 10^7$ dumbbells uniformly placed inside the droplet; the flow is continued until dimensionless time $t^* = t\dot{\gamma} = 10$, with a small time step to accurately solve the internal configurations of the dumbbells ($dt = 1/200$), taking $N_t = 2000$ time steps to finish each simulation; the number of marker particles to improve the definition of the interface was $N_{mp} = 2.5 \cdot 10^5$. The ratio between viscous and inertial effects in the system, represented by the Reynolds number, are of utmost importance, in the sense that, if the method is not able to deal with extremely low Re ("creeping" or "Stokes" flows), the inertial effects become relevant when small time steps are used, thus affecting the history of the flow and, consequently, that of the polymer particles (dumbbells); it is for this reason that the value of $Re = 0.1$, deemed appropriate in the previous section for Newtonian flows, is now replaced by $Re = 10^{-5}$ to suppress any undesired effects that inertia may have on the computation of the steady-state values for the polymer solution. The remaining dimensionless parameters are chosen as those found in [21], with $Fr \to \infty$, $We = 10^{-6}$, so that the Capillary number $Ca = 0.1$; our concentration parameter c, according to the characteristic scales chosen in [21], corresponds to $c = 1 - \beta$, with $\beta = 0.5$ the retardation parameter of the Oldroyd-B fluid; the Deborah numbers studied are $De = \{0.25; 0.5; 1; 2\}$; and the density and viscosity ratios between the drop and the outer ambient fluid (matrix) are taken as $\rho_2/\rho_1 = \mu_2/\mu_1 = 1$. We performed a set of simulations to obtain the evolution of the drop deformation and compared the results found in Figure 1 of [21].

Despite the rather coarse mesh used (equivalent to a uniform size $h = 1/40$), the not-so high number of dumbbells and the totally different approach taken by the two techniques compared—the diffuse-interface method combined with a phase-field approach ruled by Cahn–Hilliard dynamics to study the Newtonian/non-Newtonian problem in a unified way of [21] and our FEM-based, stochastic method—the results presented in Figure 5 are in remarkably good agreement, especially the steady-state values of the deformation parameter D. However, also the transient behavior shows a noteworthy resemblance, with the overshoot appearing for sufficiently high De values and the evolution of $De = 2$ being for $t \lesssim 4$ higher than those for $De = 1$. In any case, we notice the effect of the drop viscoelasticity as a means to reduce the deformation of the interface; plots of the actual shape of the interface (not included here for brevity's sake) show this same trend. Apart from the stochastic noise, the modification of the interface by the correction step of the marker particles and the mass conservation step add to the oscillatory behavior of D observed in Figure 5, which is explicitly computed from the discrete interface. Moreover, Figure 5 underscores the elastic effects of the drop, since an increase in the Deborah number De produces a larger ratio between the maximum and the steady-state value of the deformation parameter D, as a consequence of the longer time (slower response) that the polymer molecules take to recover from the applied shear strain.

We now compare the steady-state values of the polymer stress tensor obtained using the Oldroyd-B constitutive model of [22] and the Hookean dumbbell model. The results are presented in Figure 6. As we did for the Newtonian/Newtonian case, we represent the interface also in polar coordinates with $\theta \in [0, \pi]$. The normal component of the polymer stress tensor $\tau_{p,n} \equiv \mathbf{n} \cdot \boldsymbol{\tau}_p \cdot \mathbf{n}$ (with $\mathbf{n}$ the outer normal at each point of the interface) is computed in Figure 6a.

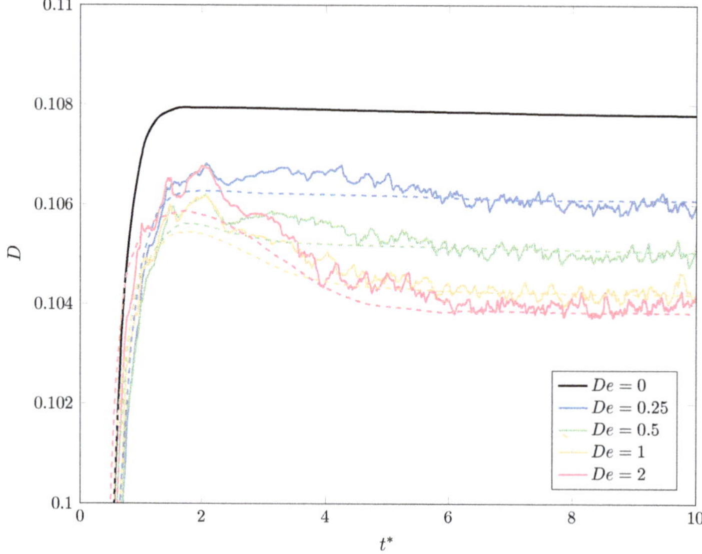

Figure 5. Comparison of the evolution of deformation for a non-Newtonian (Oldroyd-B) drop in a Newtonian matrix of equal viscosity after the abrupt start of shear flow, for increasing values of the Deborah number *De*, between our results (solid line) and those found in Yue et al. [21] (dashed line).

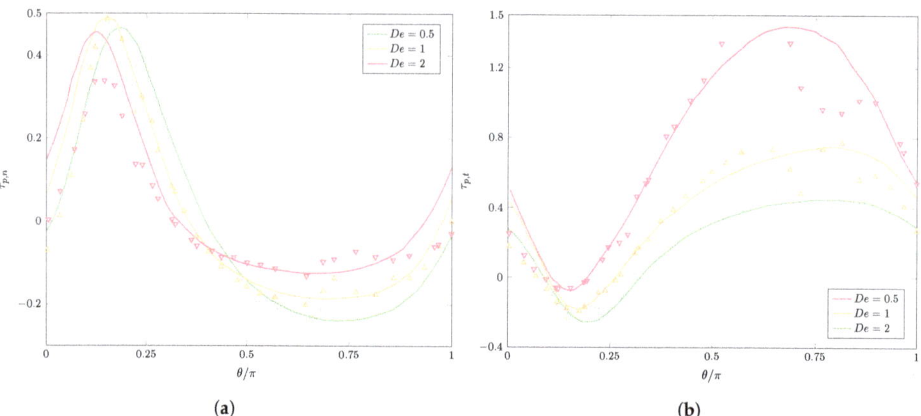

(a)   (b)

Figure 6. Comparison of the (a) normal and (b) tangential stresses along the inner edge of the interface for a non-Newtonian (Oldroyd-B) drop in a Newtonian matrix of equal viscosity, for increasing values of the Deborah number, between our results ($De = 0.5$, green squares; $De = 1$, yellow triangles; $De = 2$, inverted red triangles) and those found in Yue et al. [22].

In spite of the largely dissimilar methods, we notice a very good agreement between them. We observe how the effects of the viscoelastic drop are not very strong on the normal component of the polymer stress tensor, though increasing the viscoelasticity by means of *De* reduces the maximum value, which is also moved towards smaller orientation angles, thus playing a part in diminishing drop deformation. Overall, the normal polymer stress component is weakened at the poles and at the equator of the drop as viscoelasticity increases, preventing drop deformation.

Figure 6b shows the tangential component of the polymer stress tensor $\tau_{p,t} \equiv \mathbf{t} \cdot \boldsymbol{\tau}_p \cdot \mathbf{t}$, with $\mathbf{t}$ the unit vector tangent at all points to the interface and perpendicular to $\mathbf{n}$ (so that, in 2D, $t_x = n_y, t_y = -n_x$). Here, the polymer stresses are much larger than for the normal component, showing again an excellent agreement between our results and those in [22]. The larger De values intensify the tangential stresses at the equator, being responsible to a larger extent for the reduced deformation observed for the higher De. Despite small discrepancies, even with a coarse mesh and computationally inexpensive simulation, our results compare very well with those of the reference, capturing the minimum and maximum values of $\tau_{p,t}$, as well as the evolution of $\tau_{p,t}$ along the interface.

3.2. Drop Deformation in Buoyancy-Driven Flow

After studying the problem of drop deformation in unsteady, shear flow, in this section, we focus on following the behavior of a Newtonian drop immersed in an ambient, polymeric solution, rising due to buoyancy effects and the density ratio between the fluids [75]. By means of a series of numerical experiments, we gain insight into the deformation of the drop shape, characterized by the circularity ϕ, the ratio between the perimeter of a circle whose area is equal to that of the drop and the perimeter of the drop. We also shed light on the response of the "extra-" stress tensor $\boldsymbol{\tau}_p$ to a varying degree of polymer concentration and relaxation time, when the drop undergoes large deformation and velocity gradients, highlighting the influence of different smoothness in the CSRBF. Finally, we observe the effects that viscoelasticity may have on the flow, with emphasis on the flow streamlines during the full, unsteady simulation.

Thus, we consider a dimensionless, 2D rectangular domain $[0,1] \times [0,2]$ in which a drop of radius $R = 0.25$ is placed at position $(0.5, 0.5)$ inside an unstructured mesh of uniform size h. The outer, non-Newtonian polymer solution is modeled using the FENE and Hookean dumbbell models. The flow is further defined by the density and viscosity ratios $\rho_2/\rho_1, \mu_2/\mu_1$ and dimensionless numbers Re, We, Fr, c, De. The "no-slip" boundary condition is applied at the top and at the bottom, while we consider "free-slip" at the lateral boundaries. All simulations are carried until dimensionless time $t = 3$.

3.2.1. Convergence Results

We turn our attention to the evolution of the drop shape during the full simulation, as the mesh size is decreased from $h = 1/40$ down to $h = 1/320$, and the number of ensembles scattered over the domain changes, using two different kinetic models, Hooke, and FENE, and different degrees of viscoelasticity, $(c = 1, De = 1)$ and $(c = 5, De = 3)$. The other relevant parameters in this case are: $Re = 35, We = 10, Fr = 1, \rho_2/\rho_1 = 10^{-1} = \mu_2/\mu_1$.

In Figure 7, we present the results for convergence under mesh refinement, for the FENE and Hooke models: for the case with weaker viscoelastic effects $(c = 1, De = 1)$, the number of ensembles is $N_{ens} = 5000$, while the number of dumbbells inside each ensemble is $N_d = 150,000$ for both stochastic models. When the polymer concentration is increased $(c = 5, De = 3)$, we scatter more ensembles in the ambient (polymeric) fluid $N_{ens} = 5000$ to capture the more dramatic effects in shape deformation, reducing the number of dumbbells within each ensemble to $N_d = 15,000$ to keep the computational demands at a similar level. The results obtained here compare very well with those provided in [7] for purely uncorrelated dumbbells, showing excellent convergence with mesh refinement for both kinetic models and degrees of polymer concentration. Only in the coarser mesh, $h = 1/40$, do we observe a slightly different behavior to that produced by the other unstructured meshes. This pattern is more evident when using the Hooke instead of the FENE dumbbell model, due to the larger stresses produced by the former model during drop deformation: the minimum value of the circularity decreases for both $c = 1$ and $c = 5$, and we observe a lower local maximum at $t \approx 2.2$ for $c = 5, De = 3$, with the drop showing a lack of deformation at the latter stages of the simulation $(t \geq 2.75)$.

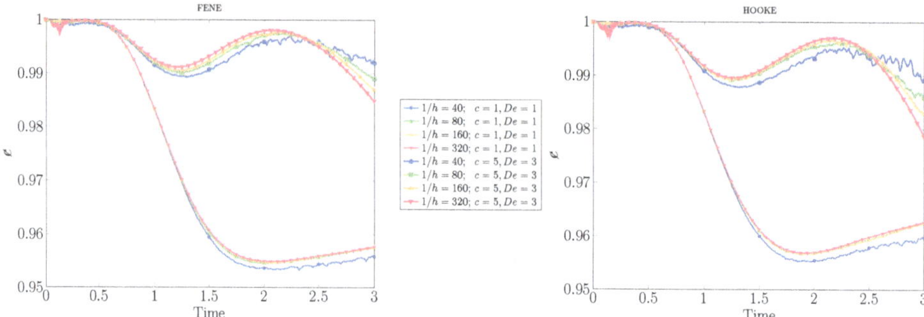

Figure 7. Evolution of the circularity of a drop rising due to buoyancy effects in a polymer solution, using the FENE and Hookean kinetic models, variable mesh size $h = \{1/40, 1/80, 1/160, 1/320\}$, and different degrees of viscoelasticity: $c = 1, De = 1$; and $c = 5, De = 3$.

Figure 8 shows the evolution of the circularity when the number of ensembles N_{ens} and dumbbells per ensemble N_d changes, keeping the product $N_{ens} \times N_d = 750 \times 10^6$ constant so as to maintain the computational cost; the mesh size is $h = 1/320$. For the case with lower polymer concentration and viscoelastic effects, the convergence results are remarkable for both models. When the concentration parameter and Deborah number are increased to $c = 5, De = 3$, we observe a small discrepancy in the circularity values, proving that a certain number of ensembles are required to reconstruct, in an accurate way, the stronger polymer stresses. However, there seems to be an optimal value of N_{ens} and N_d for a given computational cost, with this value of N_{ens} being larger for Hooke than for the FENE model. In any case, such a modest diverging behavior is only observed at or after the minimum circularity ($c = 1$) or local maximum circularity ($c = 5$) are attained.

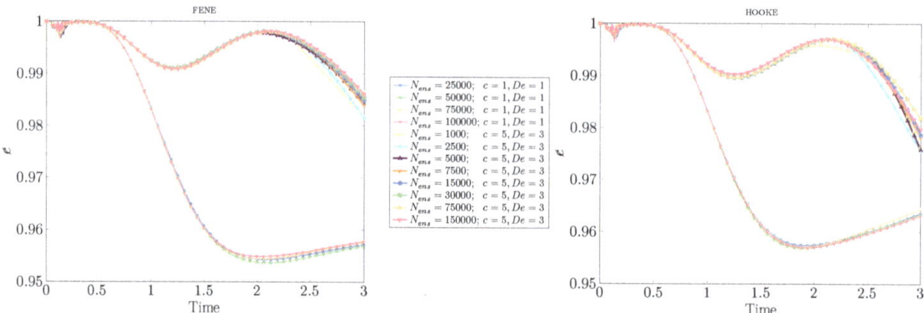

Figure 8. Evolution of the circularity of a drop rising due to buoyancy effects in a polymer solution, using the FENE and Hookean kinetic models, variable number of ensembles ($N_{ens} \times N_d = 750 \times 10^6$), and different degrees of viscoelasticity: $c = 1, De = 1$; and $c = 5, De = 3$.

3.2.2. Impact of CSRBF smoothness on the polymer stress tensor

Next, we investigate the influence that the degree of CSRBF smoothness has on the flow. In particular, we are interested in observing qualitative and quantitative deviations in the shear component and normal stress difference of the polymer stress tensor τ_p, in a case with strong viscoelastic effects ($c = 5, De = 3$), when Wendland's CSRBFs, $\varphi_{3,k}, \forall j = 0, .., 3$, are used. The reason behind our pursuing this investigation is twofold: since the "extra-"stress is responsible for the viscoelastic effects, it is mandatory to represent it as accurately as possible in simulations that aim to highlight the non-Newtonian behavior of the polymer solution; at the same time, this should be accomplished as efficiently as possible, opting for the CSRBFs that provide results nearly as accurate as the more computationally expensive alternatives.

In the following simulations, we take an unstructured mesh of size $h = 1/320$, keeping the same values of the dimensionless parameters found in the previous section. The number of ensembles for the FENE model is $N_{ens} = 30,000$, and the number of dumbbells within each ensemble $N_d = 25,000$; for the numerical experiments using the Hookean dumbbell model, the parameters are $N_{ens} = 75,000$ and $N_d = 10,000$. Thus, Figures 9–12 show the shear component $\tau_{p,12}$ and the normal stress difference $\tau_{p,11} - \tau_{p,22}$, at (dimensionless) instants of time $t = 2$ and $t = 3$; for the ease of representation, the Hookean stresses are scaled down a factor of 1.81 from the results obtained with the FENE model.

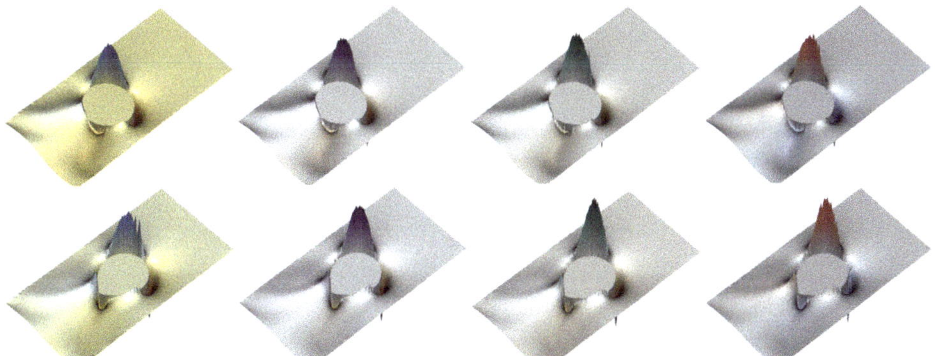

Figure 9. Representation of the shear component of the polymer stress tensor, $\tau_{p,12}$, for the FENE model ($b = 50$) with $c = 5, De = 3$, at dimensionless instants of time $t = 2$ and $t = 3$, for variable types of CSRBF. From left to right: $\varphi_{3,0}, \varphi_{3,1}, \varphi_{3,2}, \varphi_{3,3}$. Top row: $t = 2$; bottom row: $t = 3$.

For the shear component $\tau_{p,12}$ of the polymer stress tensor using the FENE model (Figure 9), all CSRBFs provide very smooth solutions at $t = 2$. At the final time of the simulation $t = 3$, when larger stresses are produced, the Wendland $\varphi_{3,0}$ CSRBF shows small oscillations and a certain roughness that may translate into an insufficiently resolved velocity field and an overall loss of symmetry in the drop shape. Wendland $\varphi_{3,1}$ offers much better accuracy, as well as an improved spatial resolution of the stresses, to such an extent that increasing the smoothness of the CSRBF ($\varphi_{3,0}$, $\varphi_{3,3}$) offers almost no distinct enhancement.

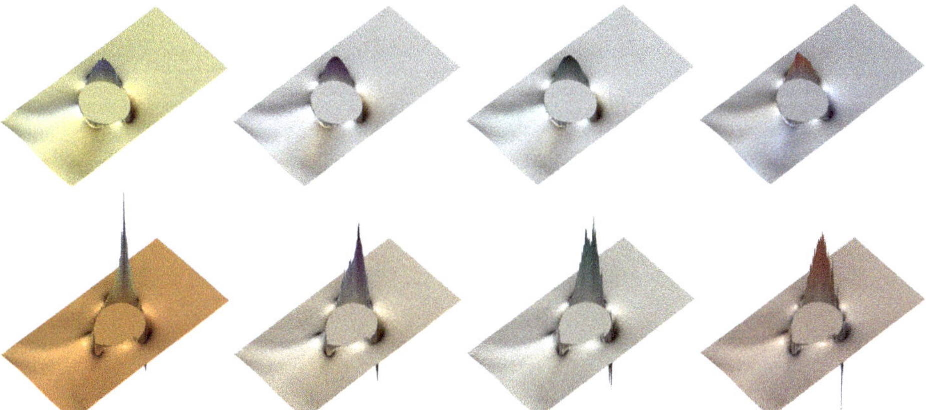

Figure 10. Representation of the shear component of the polymer stress tensor, $\tau_{p,12}$, for the Hookean dumbbell model (Oldroyd-B) with $c = 5, De = 3$, at dimensionless instants of time $t = 2$ and $t = 3$, for variable types of CSRBF. From left to right: $\varphi_{3,0}, \varphi_{3,1}, \varphi_{3,2}, \varphi_{3,3}$. Top row: $t = 2$; bottom row: $t = 3$.

When the Hookean dumbbell model is used (Figure 10), $\tau_{p,12}$ shows much larger values throughout the whole simulation. The influence of the type of CSRBF becomes evident at the later stages, when extreme and very localized stresses are observed. The less smooth $\varphi_{3,0}$ is unable to provide a symmetric solution, while the other Wendland functions offer much better results in terms of symmetry.

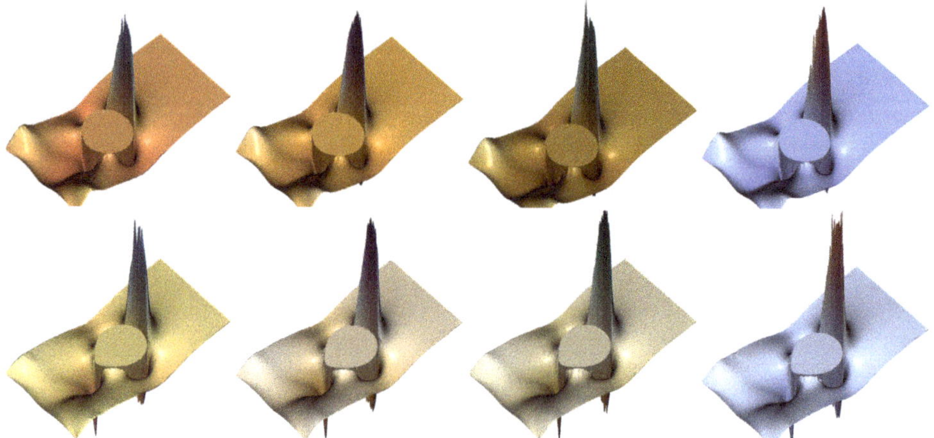

Figure 11. Representation of the normal stress difference of the polymer stress tensor, $\tau_{p,11} - \tau_{p,22}$, for the FENE model ($b = 50$) with $c = 5, De = 3$, at dimensionless instants of time $t = 2$ and $t = 3$, for variable types of CSRBF. From left to right: $\varphi_{3,0}, \varphi_{3,1}, \varphi_{3,2}, \varphi_{3,3}$. Top row: $t = 2$; bottom row: $t = 3$.

If we now inspect the results for the normal stress difference, $\tau_{p,11} - \tau_{p,22}$, we observe that when the FENE model is used (Figure 11), remarkable smoothness is retrieved by all the CSRBFs explored, but for $\varphi_{3,0}$, which somewhat underperforms at the last instants of the simulation, when the viscoelastic effects have shown up and the drop attains maximum deformation.

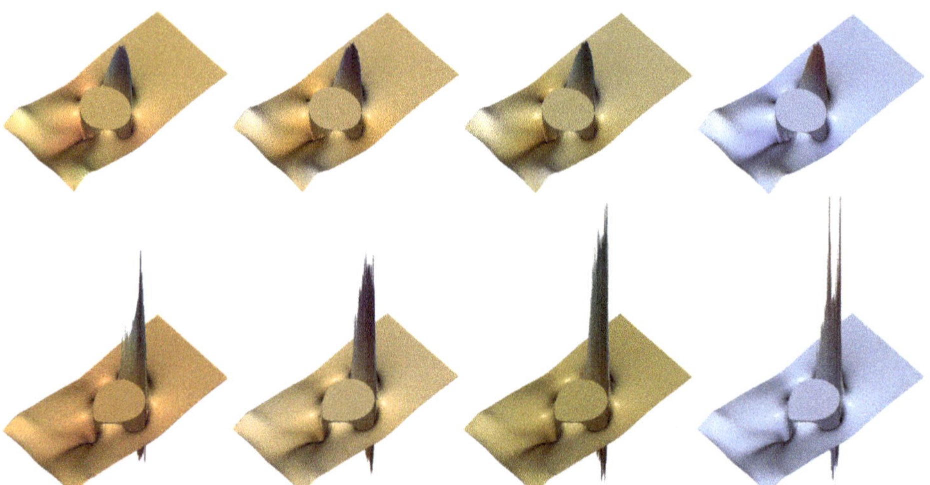

Figure 12. Representation of the normal stress difference of the polymer stress tensor, $\tau_{p,11} - \tau_{p,22}$, for the Hookean dumbbell model (Oldroyd-B) with $c = 5, De = 3$, at dimensionless instants of time $t = 2$ and $t = 3$, for variable types of CSRBF. From left to right: $\varphi_{3,0}, \varphi_{3,1}, \varphi_{3,2}, \varphi_{3,3}$. Top row: $t = 2$; bottom row: $t = 3$.

Figure 12 represents the normal stress difference, $\tau_{p,11} - \tau_{p,22}$, for the Hooke model. In this case, a higher degree of smoothness has a positive effect, improving the symmetry of the solution, to a certain extent. However, unless the spatial resolution of the velocity field and the velocity gradients is improved, the number of ensembles distributed over the domain is large enough, and the dumbbells per ensemble ensure a sufficiently accurate computation of the "extra-"stress at the ensembles, there is no point in further increasing the quality of the CSRBF for polymer stress reconstruction, since the extremely large, very localized stresses cannot be accurately computed unless the previous conditions are met. On the other hand, all previous conditions can be satisfied so long as the necessary computational resources are available. As a result of all the previous investigations, we find Wendland's $\varphi_{3,1}$ CSRBF as a good choice in terms of accuracy, smoothness, and computational demands, used in this work unless otherwise noted.

3.2.3. Flow Pattern under Increasing Viscoelastic Effects

Finally, we explore the flow pattern, streamlines, and isocontours of the polymer stress tensor, in a series of simulations under increasing polymer concentration and relaxation times, using the FENE dumbbell model with extensibility parameter $b = 50$; see Figures 13–16. For all the subsequent numerical experiments, a uniform, unstructured mesh of size $h = 1/320$ is used, with the number of dumbbells per ensemble N_d and the number of ensembles N_{ens} being collected in Table 1. Notice that, for stronger viscoelastic effects, a larger number of ensembles are scattered throughout the domain, reducing the computationally available number of dumbbells per ensemble, which in turn translates into a larger stochastic noise. Two different density and viscosity ratios are explored to underscore the effects caused by the polymer solution on the shape deformation and flow pattern of the drop $(\rho_2/\rho_1 = 10^{-1} = \mu_2/\mu_1, Re = 35, We = 10, Fr = 1)$ and of the bubble $(\rho_2/\rho_1 = 10^{-3}, \mu_2/\mu_1 = 10^{-2}, Re = 35, We = 125, Fr = 1)$.

Table 1. Each entry in the table shows N_d / N_{ens} for increasing viscoelastic effects and different density ratios. N_d is the number of dumbbells per ensemble and N_{ens} the number of ensembles scattered in the domain.

Density Ratio	$c = 1, De = 1$	$c = 3, De = 1$	$c = 5, De = 3$	$c = 9, De = 5$
10	150,000 / 5000	75,000 / 10,000	50,000 / 15,000	37,500 / 20,000
1000	150,000 / 5000	50,000 / 15,000	15,000 / 50,000	15,000 / 5000

In Figure 13, we show the final shape and filled isocontours of the normal stress difference of the polymer stress tensor for increasing concentration of the polymer solution and longer relaxation times for the drop (density ratio of 10), while Figure 14 presents the behavior of the shear component under the same circumstances, for the bubble immersed in the polymeric media (density ratio 1000). For the drop and normal stress difference, the results present an excellent degree of symmetry and smoothness, despite the underlying stochastic procedure, even for the strongest viscoelastic effects at $(c = 5, De = 3)$, which produce extreme values at the top and bottom of the rising drop.

The bubble depicted in Figure 14 with the shear stress isocontours shows how the harder computational demands of this simulation translate into a modest reduction of the overall symmetry along with more dramatic changes of the bubble shape. Nevertheless, the smoothness of the discrete solution obtained for the polymer stress tensor, even when the stronger polymer concentration is used, is such that it is possible to retrieve the maximum values at the wake of the bubble, with a high degree of accuracy and symmetry. Figures 13 and 14 also evince elastic effects, with heightened drop deformation caused by larger values of the Deborah number (i.e., relaxation time of the polymer) and polymer concentration, while viscous and surface tension forces are kept at the same level.

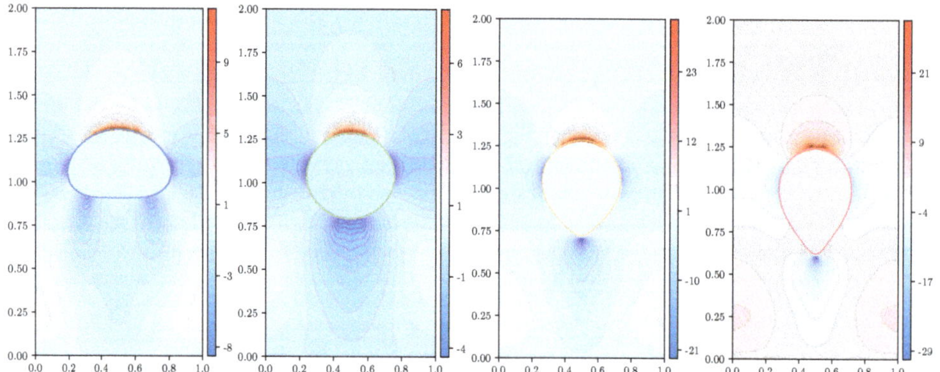

Figure 13. Drop shape and isocontours of normal stress difference of the polymer $\tau_{p,11} - \tau_{p,22}$ for the FENE model ($b = 50$) with $\rho_2/\rho_1 = 10^{-1} = \mu_2/\mu_1$, at $t = 3$. From left to right, $c = 1, De = 1$; $c = 3, De = 1$; $c = 5, De = 3$; and $c = 9, De = 5$.

Next, we focus on the streamlines of the flow under increasing viscoelastic effects. In Figure 15, we represent the flow pattern for the drop (moderate density and viscosity ratios), with Figure 15a presenting a snapshot at time $t = 2$, while Figure 15b shows the pattern at the final instant of time, $t = 3$. For these high values of polymer concentration and relaxation times, at $t = 3$, we observe the downwards velocities that characterize the "negative wake" effect [23,76].

For large density and viscosity ratios, Figure 16 depicts the behavior of the deforming bubble at $t = 3$. At the higher levels of polymer concentration and relaxation times, the area in which downwards velocities are observed, comprising the "negative wake", is much larger than in the case of the deforming drop.

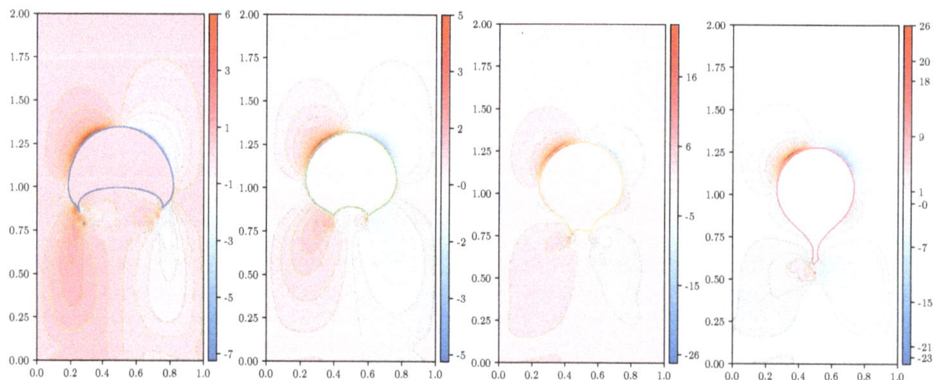

Figure 14. Drop shape and shear component of the polymer stress tensor $\tau_{p,12}$ for the FENE model ($b = 50$) with $\rho_2/\rho_1 = 10^{-3}, \mu_2/\mu_1 = 10^{-2}$, at $t = 3$. From left to right, $c = 1, De = 1$; $c = 3, De = 1$; $c = 5, De = 3$; and $c = 9, De = 5$.

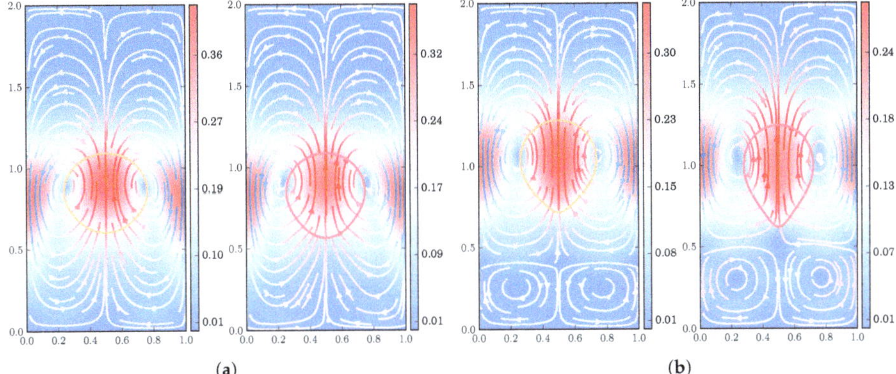

(a) (b)

Figure 15. Drop shape and streamlines for the FENE model ($b = 50$), with $\rho_2/\rho_1 = 10^{-1} = \mu_2/\mu_1$ at different instants of time: (**a**) $t = 2$ and (**b**) $t = 3$. For each panel, the left figure (yellow line) shows the streamlines for $c = 5, De = 3$; the right figure (red line), those for $c = 9, De = 5$.

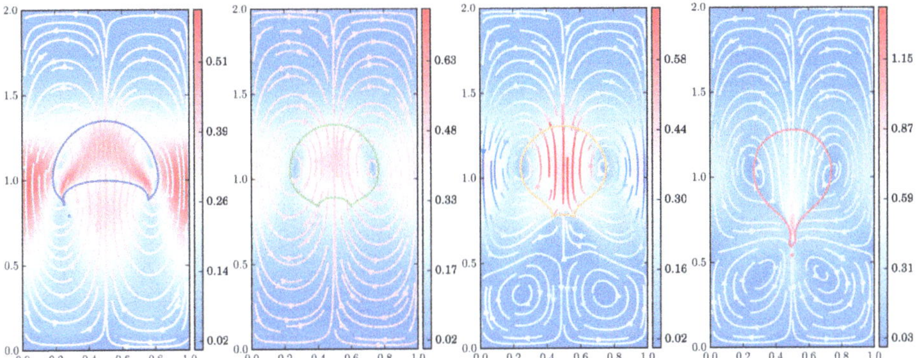

Figure 16. Drop shape and streamlines for the FENE model ($b = 50$) with $\rho_2/\rho_1 = 10^{-3}, \mu_2/\mu_1 = 10^{-2}$, at $t = 3$. From left to right, $c = 1, De = 1$ (blue line); $c = 3, De = 1$ (green line); $c = 5, De = 3$ (yellow line); and $c = 9, De = 5$ (red line).

4. Conclusions

This paper presents a series of numerical experiments in an effort to gain insight into the impact that viscoelasticity may have on drop deformation, under a number of multiphase flow configurations. The numerical method employed is based on an FEM-based spatial discretization, using a semi-Lagrangian approach to deal with the convective terms, a kinetic modeling of the polymer contribution to the stress tensor, and ML-inspired techniques for building, over the whole domain, each of the components of the extra-stress tensor, which effectively decouples the "microscopic" and "macroscopic" scales. Hence, the process of mesh refinement is no longer hampered by the low number of polymer particles at a certain computational cell, allowing us to use very refined meshes and achieve excellent smoothness and accuracy in kinetic-based, complex flow simulations.

The results on drop deformation under shear flow agree extremely well with those in the literature produced by an equivalent constitutive (Oldroyd-B) model and indicate that drop viscoelasticity prevents deformation to some degree, underscoring the efficiency and viability of the stochastic approach in multiphase flows. For more demanding situations in buoyancy-driven flow, we obtain remarkable mesh (macro) and ensemble (micro) convergence for both the FENE and Hooke models; as for the degree of CSRBFs smoothness required to accurately build the "extra-"stress tensor, the numerical study points to a minimum degree of C^2 smoothness guaranteed by $\varphi_{3,1}$, with higher smoothness providing small benefits compared to the additional computational cost. Finally, we observe dramatic changes of flow pattern, including viscoelastic ("negative-wake") effects and extremely large values of shear and normal stress differences close to the drop interface, as the polymer concentration and relaxation times increase.

In future investigations of multiphase flow of polymeric liquids, we will try to address improved resolutions and problem configurations taking advantage of this ML-enhanced method to combine isotropic and anisotropic mesh adaptation [77] with a kinetic modeling approach.

Funding: This research was funded by "Ministerio de Ciencia, Innovación y Universidades" under Grant PGC2018-097565-B-I00.

Conflicts of Interest: The author declares no conflict of interest.

Abbreviations The following abbreviations are used in this manuscript:

μ_1	Viscosity of the continuous phase
μ_2	Droplet viscosity
ρ_1	Density of the continuous phase
ρ_2	Droplet density
τ_p	Polymer ("extra-")stress tensor
$\tau_{p,12}$	Shear component of the polymer stress tensor
$\tau_{p,11-22}$	Normal stress difference of the polymer stress tensor
ϕ	Level set function
φ	Compactly-supported Wendland function
χ	Support size of the CSRBF
ψ	Trial basis function
b	FENE extensibility parameter
c	Concentration parameter
$\cancel{c}$	Droplet circularity
dt	Time step size
h	Grid size of the uniform, unstructured mesh
p	Pressure field
$\mathbf{v}$	Velocity field
s	Approximation interpolant of the CSRBF
D	Deformation parameter
N_d	Number of uncorrelated dumbbells per ensemble
N_{ens}	Number of ensembles of polymer particles
N_{mp}	Number of marker particles
Ca	Capillary number
Fr	Froude number
Re	Reynolds number
We	Weber number
ALE	Arbitrary Lagrangian-Eulerian method
BD	Brownian Dynamics simulations
CSRBF	Compactly-Supported Radial Basis Function
FEM	Finite Element Method
FENE	Finitely Extensible Non-linear Elastic model
LS	Level Set method
LSC	Least Squares Commutator preconditioner
ML	Machine Learning
NN	Nearest Neighbor
PBC	Periodic Boundary Conditions
PLS	Particle Level Set
RBF	Radial Basis Function
VOF	Volume-Of-Fluid method
[1 c]	Vector of interpolation coefficients for the CSRBF
K	Discrete matrix system
PETSc	Portable, Extensible Toolkit for Scientific Computation

References

1. Zenit, R.; Feng, J. Hydrodynamic Interactions among Bubbles, Drops, and Particles in Non-Newtonian Liquids. *Annu. Rev. Fluid Mech.* **2018**, *50*, 505–534.
2. Aytouna, M.; Paredes, J.; Shahidzadeh-Bonn, N.; Moulinet, S.; Wagner, C.; Amarouchene, Y.; Eggers, J.; Bonn, D. Drop Formation in Non-Newtonian Fluids. *Phys. Rev. Lett.* **2013**, *110*, 034501.
3. Prieto-López, L.O.; Xu, J.; Cui, J. Magnetic-Responsive Bendable Nozzles for Open Surface Droplet Manipulation. *Polymers* **2019**, *11*, 1792.
4. Feigl, K.; Baniabedalruhman, A.; Tanner, F.X.; Windhab, E.J. Numerical simulations of the breakup of emulsion droplets inside a spraying nozzle. *Phys. Fluids* **2016**, *28*, 123103. doi:10.1063/1.4972097.
5. Fortelný, I.; Jůza, J. Description of the Droplet Size Evolution in Flowing Immiscible Polymer Blends. *Polymers* **2019**, *11*, 761.
6. Mohammad Karim, A. Experimental dynamics of Newtonian and non-Newtonian droplets impacting liquid surface with different rheology. *Phys. Fluids* **2020**, *32*, 043102.
7. Prieto, J.L. Stochastic particle level set simulations of buoyancy-driven droplets in non-Newtonian fluids. *J. Non-Newtonian Fluid Mech.* **2015**, *226*, 16–31.
8. Faulde, M.; Siemes, E.; Wöll, D.; Jupke, A. Fluid Dynamics of Microgel-Covered Drops Reveal Impact on Interfacial Conditions. *Polymers* **2018**, *10*, 809.
9. Elgeti, S.; Sauerland, H. Deforming fluid domains within the finite element method: Five mesh-based tracking methods in comparison. *Arch. Comput. Method. E.* **2016**, *23*, 323–361.
10. Donea, J.; Huerta, A.; Ponthot, J.P.; Rodriguez-Ferran, A. *Encyclopedia of Computational Mechanics Vol. 1: Fundamentals., Chapter 14: Arbitrary Lagrangian-Eulerian Methods*; Wiley & Sons: Chichester, UK, 2004.
11. Bo, W.; Shashkov, M. Adaptive reconnection-based arbitrary Lagrangian Eulerian method. *J. Comput. Phys.* **2015**, *299*, 902–939.
12. Kenamond, M.; Shashkov, M. The distribution-based remapping of the nodal mass and momentum between arbitrary meshes for staggered arbitrary Lagrangian-Eulerian hydrodynamics. *Comput. Fluids* **2020**, *201*, 104469.
13. Hirt, C.W.; Nichols, B.D. Volume of fluid (VOF) method for the dynamics of free boundaries. *J. Comput. Phys.* **1981**, *39*, 201–225.
14. López, J.; Hernández, J.; Gómez, P.; Faura, F. A volume of fluid method based on multidimensional advection and spline interface reconstruction. *J. Comput. Phys.* **2004**, *195*, 718–742.
15. López, J.; Hernández, J.; Gómez, P.; Faura, F. Non-convex analytical and geometrical tools for volume truncation, initialization and conservation enforcement in VOF methods. *J. Comput. Phys.* **2019**, *392*, 666–693.
16. Sethian, J.A.; Smereka, P. Level set methods for fluid interfaces. *Annu. Rev. Fluid Mech.* **2003**, *35*, 341–372.
17. Gibou, F.; Fedkiw, R.; Osher, S. A review of level-set methods and some recent applications. *J. Comput. Phys.* **2018**, *353*, 82–109.
18. Enright, D.; Fedkiw, R.; Ferziger, J.; Mitchell, I. A Hybrid Particle Level Set Method for Improved Interface Capturing. *J. Comput. Phys.* **2002**, *183*, 83–116.
19. Thome, J.R. *Encyclopedia of Two-Phase Heat Transfer and Flow III*; World Scientific Publishing: Singapore, 2018. doi:10.1142/10656.
20. Yue, P.; Feng, J.J.; Liu, C.; Shen, J. A diffuse-interface method for two-phase complex fluids. *J. Fluid Mech.* **2004**, *515*, 293–317.
21. Yue, P.; Feng, J.J.; Liu, C.; Shen, J. Transient drop deformation upon startup of shear in viscoelastic fluids. *Phys. Fluids* **2005**, *17*, 123101.
22. Yue, P.; Feng, J.J.; Liu, C.; Shen, J. Viscoelastic effects on drop deformation in steady shear. *J. Fluid Mech.* **2005**, *540*, 427–437.
23. Pillapakkam, S.B.; Singh, P.; Blackmore, D.; Aubry, N. Transient and steady state of a rising bubble in a viscoelastic fluid. *J. Fluid Mech.* **2007**, *589*, 215–252.
24. Foteinopoulou, K.; Laso, M. Numerical simulation of bubble dynamics in a Phan-Thien-Tanner liquid: Non-linear shape and size oscillatory response under periodic pressure. *Ultrasonics* **2010**, *50*, 758–776.
25. Castillo, E.; Baiges, J.; Codina, R. Approximation of the two-fluid flow problem for viscoelastic fluids using the level set method and pressure enriched finite element shape functions. *J. Non-Newtonian Fluid Mech.* **2015**, *225*, 37–53. doi:10.1016/j.jnnfm.2015.09.004.

26. Fraggedakis, D.; Pavlidis, M.; Dimakopoulos, Y.; Tsamopoulos, J. On the velocity discontinuity at a critical volume of a bubble rising in a viscoelastic fluid. *J. Fluid Mech.* **2016**, *789*, 310–346.

27. Wang, Y.; Wang, Y.; Cheng, Z. Direct Numerical Simulation of Gas-Liquid Drag-Reducing Cavity Flow by the VOSET Method. *Polymers* **2019**, *11*, 596. doi:10.3390/polym11040596.

28. Xie, C.; Xu, K.; Mohanty, K.; Wang, M.; Balhoff, M.T. Nonwetting droplet oscillation and displacement by viscoelastic fluids. *Phys. Rev. Fluids* **2020**, *5*, 063301.

29. Keunings, R. Micro-macro methods for the multiscale simulation of viscoelastic flow using molecular models of kinetic theory. In *Rheology Reviews*; Binding, D.M.; Walters, K., Eds.; British Society of Rheology: Aberystwyth, UK, 2004; pp. 67–98.

30. Degond, P.; Lozinski, A.; Owens, R.G. Kinetic models for dilute solutions of dumbbells in non-homogeneous flows revisited. *J. Non-Newtonian Fluid Mech.* **2010**, *165*, 509–518.

31. Müller, M.; de Pablo, J.J. Computational Approaches for the Dynamics of Structure Formation in Self-Assembling Polymeric Materials. *Annu. Rev. Mater. Res.* **2013**, *43*, 1–34.

32. Stephanou, P.; Kröger, M. Tumbling-Snake Model for Polymeric Liquids Subjected to Biaxial Elongational Flows with a Focus on Planar Elongation. *Polymers* **2018**, *10*, 329.

33. Megariotis, G.; Vogiatzis, G.; Sgouros, A.; Theodorou, D. Slip Spring-Based Mesoscopic Simulations of Polymer Networks: Methodology and the Corresponding Computational Code. *Polymers* **2018**, *10*, 1156.

34. Laso, M.; Öttinger, H.C. Calculation of viscoelastic flow using molecular models: The CONNFFESSIT approach. *J. Non-Newtonian Fluid Mech.* **1993**, *47*, 1–20.

35. Cormenzana, J.; Ledda, A.; Laso, M.; Debbaut, B. Calculation of free surface flows using CONNFFESSIT. *J. Rheol.* **2001**, *45*, 237–258.

36. Grande, E.; Laso, M.; Picasso, M. Calculation of variable-topology free surface flows using CONNFFESSIT. *J. Non-Newtonian Fluid Mech.* **2003**, *113*, 127–145.

37. Prieto, J.L. SLEIPNNIR: A multiscale, particle level set method for Newtonian and non-Newtonian interface flows. *Comput. Methods Appl. Mech. Eng.* **2016**, *307*, 164–192.

38. Prieto, J.L.; Bermejo, R.; Laso, M. A semi-Lagrangian micro-macro method for viscoelastic flow calculations. *J. Non-Newtonian Fluid Mech.* **2010**, *165*, 120–135.

39. Bajaj, M.; Bhat, P.P.; Prakash, J.R.; Pasquali, M. Multiscale simulation of viscoelastic free surface flows. *J. Non-Newtonian Fluid Mech.* **2006**, *140*, 87–107.

40. Xu, X.; Ouyang, J.; Li, W.; Liu, Q. SPH simulations of 2D transient viscoelastic flows using Brownian Configuration Fields. *J. Non-Newtonian Fluid Mech.* **2014**, *208–209*, 59–71.

41. Hulsen, M.A.; van Heel, A.P.G.; van den Brule, B.H.A.A. Simulation of viscoelastic flows using Brownian Configuration Fields. *J. Non-Newtonian Fluid Mech.* **1997**, *70*, 79–101.

42. Shalev-Shwartz, S.; Ben-David, S. *Understanding Machine Learning: From Theory to Algorithms*; Cambridge University Press: New york, NY, USA, 2014.

43. Goodfellow, I.; Bengio, Y.; Courville, A. *Deep Learning*; The MIT Press: Cambridge, MA, USA, 2016.

44. Doblies, A.; Boll, B.; Fiedler, B. Prediction of Thermal Exposure and Mechanical Behavior of Epoxy Resin Using Artificial Neural Networks and Fourier Transform Infrared Spectroscopy. *Polymers* **2019**, *11*, 363.

45. Jackson, N.E.; Bowen, A.S.; de Pablo, J.J. Efficient Multiscale Optoelectronic Prediction for Conjugated Polymers. *Macromolecules* **2020**, *53*, 482–490.

46. Kopal, I.; Harničárová, M.; Valíček, J.; Krmela, J.; Lukáč, O. Radial Basis Function Neural Network-Based Modeling of the Dynamic Thermo-Mechanical Response and Damping Behavior of Thermoplastic Elastomer Systems. *Polymers* **2019**, *11*, 1074.

47. Chandrasekaran, A.; Kim, C.; Venkatram, S.; Ramprasad, R. A Deep Learning Solvent-Selection Paradigm Powered by a Massive Solvent/Nonsolvent Database for Polymers. *Macromolecules* **2020**, *53*, 4764–4769.

48. Ma, M.; Lu, J.; Tryggvason, G. Using statistical learning to close two-fluid multiphase flow equations for a simple bubbly system. *Phys. Fluids* **2015**, *27*, 092101.

49. Ladický, L.; Jeong, S.; Solenthaler, B.; Pollefeys, M.; Gross, M. Data-Driven Fluid Simulations Using Regression Forests. *ACM Trans. Graph.* **2015**, *34*, 1–9.

50. Gibou, F.; Hyde, D.; Fedkiw, R. Sharp interface approaches and deep learning techniques for multiphase flows. *J. Comput. Phys.* **2019**, *380*, 442–463.

51. Öttinger, H.C. *Stochastic Processes in Polymeric Fluids: Tools and Examples for Developing Simulation Algorithms*; Springer: Berlin, Germany, 1996.

52. Prieto, J.L. An RBF-reconstructed, polymer stress tensor for stochastic, particle-based simulations of non-Newtonian, multiphase flows. *J. Non-Newtonian Fluid Mech.* **2016**, *227*, 90–99.

53. Bermejo, R.; Prieto, J.L. A Semi-Lagrangian Particle Level Set Finite Element Method for Interface Problems. *SIAM J. Sci. Comput.* **2013**, *35*, A1815–A1846.

54. Cheng, L.T.; Tsai, Y.H. Redistancing by flow of time dependent eikonal equation. *J. Comput. Phys.* **2008**, *227*, 4002–4017.

55. Hunter, J.D. Matplotlib: A 2D graphics environment. *Comput. Sci. Eng.* **2007**, *9*, 90–95. doi:10.1109/MCSE.2007.55.

56. Tantau, T. The TikZ and PGF Packages. Comprehensive TEX Archive Network, CTAN. 2020 Available online: http://mirrors.ctan.org/graphics/pgf/base/doc/pgfmanual.pdf (accesed on 24 July 2020).

57. Ramachandran, P.; Varoquaux, G. Mayavi: 3D Visualization of Scientific Data. *Comput. Sci. Eng.* **2011**, *13*, 40–51.

58. Pandas Development Team. Pandas-Dev/Pandas: Pandas. 2020. Available online: https://doi.org/10.5281/zenodo.3509134 (accesed on 24 July 2020).

59. Schaback, R.; Wendland, H. Kernel techniques: From machine learning to meshless methods. *Acta Numerica* **2006**, *15*, 543–639. doi:10.1017/S0962492906270016.

60. Carr, J.C.; Beatson, R.K.; Cherrie, J.B.; Mitchell, T.J.; Fright, W.R.; McCallum, B.C.; Evans, T.R. Reconstruction and Representation of 3D Objects with Radial Basis Functions. In *SIGGRAPH'2001: Proceedings of the 28th Annual Conference on Computer Graphics and Interactive Techniques*; Association for Computing Machinery: New York, NY, USA, 2001; pp. 67–76.

61. Ohtake, Y.; Belyaev, A.; Seidel, H.P. 3D scattered data interpolation and approximation with multilevel compactly supported RBFs. *Graph. Models* **2005**, *67*, 150–165.

62. Wendland, H. *Scattered Data Approximation*; Cambridge University Press: Cambridge, UK, 2005.

63. Balay, S.; Abhyankar, S.; Adams, M.F.; Brown, J.; Brune, P.; Buschelman, K.; Dalcin, L.; Dener, A.; Eijkhout, V.; Gropp, W.D.; Karpeyev, D.; et al. PETSc Web Page. Available online: https://www.mcs.anl.gov/petsc (accesed on 24 July 2020).

64. Dean, E.J.; Glowinski, R. On some finite element methods for the numerical simulation of incompressible viscous flow. In *Incompressible Computational Fluid Dynamics*; Gunzburger, M.D.; Nicolaides, R.A., Eds.; Cambridge University Press: New York, NY, USA, 1993; pp. 109–150.

65. Sousa, F.S.; Oishi, C.M.; Buscaglia, G.C. Spurious transients of projection methods in microflow simulations. *Comput. Methods Appl. Mech. Eng.* **2015**, *285*, 659–693.

66. Elman, H.; Howle, V.; Shadid, J.; Shuttleworth, R.; Tuminaro, R. A taxonomy and comparison of parallel block multi-level preconditioners for the incompressible Navier-Stokes equations. *J. Comput. Phys.* **2008**, *227*, 1790–1808.

67. Elman, H.C.; Silvester, D.J.; Wathen, A.J. *Finite Elements and Fast Iterative Solvers: With Applications in Incompressible Fluid Dynamics*; Oxford University Press: Oxford, UK, 2005.

68. Davis, T.A.; Hager, W.W. Dynamic Supernodes in Sparse Cholesky Update/Downdate and Triangular Solves. *ACM Trans. Math. Softw.* **2009**, *35*, 27:1–27:23.

69. Golub, G.H.; Loan, C.F.V. *Matrix Computations*, 4th ed.; Johns Hopkins University Press: Baltimore, MD, USA, 2012.

70. Gee, M.; Siefert, C.; Hu, J.; Tuminaro, R.; Sala, M. *ML 5.0 Smoothed Aggregation User's Guide*; Technical Report SAND2006-2649; Sandia National Laboratories: Livermore, CA, USA, 2006.

71. Pask, J.; Klein, B.; Sterne, P.; Fong, C. Finite-element methods in electronic-structure theory. *Comput. Phys. Commun.* **2001**, *135*, 1–34.

72. Sukumar, N.; Pask, J.E. Classical and enriched finite element formulations for Bloch-periodic boundary conditions. *Int. J. Numer. Methods Eng.* **2009**, *77*, 1121–1138.

73. Zhou, H.; Pozrikidis, C. The flow of suspensions in channels: Single files of drops. *Phys. Fluids* **1993**, *5*, 311–324.

74. Afkhami, S.; Yue, P.; Renardy, Y. A comparison of viscoelastic stress wakes for two-dimensional and three-dimensional Newtonian drop deformations in a viscoelastic matrix under shear. *Phys. Fluids* **2009**, *21*, 072106.

75. Balla, M.; Kavuri, S.; Tripathi, M.K.; Sahu, K.C.; Govindarajan, R. Effect of viscosity and density ratios on two drops rising side by side. *Phys. Rev. Fluids* **2020**, *5*, 013601.

76. Hassager, O. Negative wake behind bubbles in non-Newtonian liquids. *Nature* **1979**, *279*, 402–403.
77. Prieto, J.L.; Carpio, J. A-SLEIPNNIR: A multiscale, anisotropic adaptive, particle level set framework for moving interfaces. Transport equation applications. *J. Comput. Phys.* **2019**, *377*, 89–116.

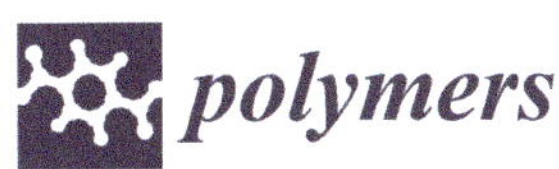 

Article

Solution Blow Spinning of High-Performance Submicron Polyvinylidene Fluoride Fibres: Computational Fluid Mechanics Modelling and Experimental Results

Rasheed Atif [1,*], Madeleine Combrinck [1], Jibran Khaliq [1], Ahmed H. Hassanin [2,3], Nader Shehata [2,4,5,6], Eman Elnabawy [2] and Islam Shyha [1]

[1] Department of Mechanical and Construction Engineering, Faculty of Engineering and Environment, Northumbria University, Newcastle upon Tyne NE1 8ST, UK; madeleine.combrinck@northumbria.ac.uk (M.C.); jibran.khaliq@northumbria.ac.uk (J.K.); islam.shyha@northumbria.ac.uk (I.S.)

[2] Centre of Smart Nanotechnology and Photonics (CSNP), SmartCI Research Center, Alexandria University, Alexandria 21544, Egypt; ahassan@ncsu.edu (A.H.H.); nader83@vt.edu (N.S.); eman.elnabawy@smartci.alexu.edu.eg (E.E.)

[3] Department of Textile Engineering, Faculty of Engineering, Alexandria University, Alexandria 21544, Egypt

[4] Department of Engineering Mathematics and Physics, Faculty of Engineering, Alexandria University, Alexandria 21544, Egypt

[5] USTAR Bioinnovations Centre, Faculty of Science, Utah State University, Logan, UT 84341, USA

[6] Kuwait College of Science and Technology (KCST), Doha District 13133, Kuwait

* Correspondence: a.rasheed@northumbria.ac.uk; Tel.: +44-(0)-191-227-3062

Received: 15 April 2020; Accepted: 8 May 2020; Published: 16 May 2020

Abstract: Computational fluid dynamics (CFD) was used to investigate characteristics of high-speed air as it is expelled from a solution blow spinning (SBS) nozzle using a k-ε turbulence model. Air velocity, pressure, temperature, turbulent kinetic energy and density contours were generated and analysed in order to achieve an optimal attenuation force for fibre production. A bespoke convergent nozzle was used to produce polyvinylidene fluoride (PVDF) fibres at air pressures between 1 and 5 bar. The nozzle comprised of four parts: a polymer solution syringe holder, an air inlet, an air chamber, and a cap that covers the air chamber. A custom-built SBS setup was used to produce PVDF submicron fibres which were consequently analysed using scanning electron microscope (SEM) for their morphological features. Both theoretical and experimental observations showed that a higher air pressure (4 bar) is more suitable to achieve thin fibres of PVDF. However, fibre diameter increased at 5 bar and intertwined ropes of fibres were also observed.

Keywords: CFD; SBS; nozzle; PVDF; fibres

---

1. Introduction

In 1969, polyvinylidene fluoride (PVDF) was first reported as thermoplastic polymer piezoelectric material (PEM) exhibiting the piezoelectric activity [1]. PVDF based PEMs are classified as stimuli responsive materials and have been employed as standalone or as matrices in composites and layered structures to fabricate stimuli responsive systems for applications such as drug delivery and tissue engineering [2–4]. One of the applications of PVDF based PEMs is intelligent clothing to sense user activities in sports and personalized health care [5–7]. As the precursor for textiles is yarn, which is produced from fibres, various fabrication methods have been employed to produce fibres, such as gas jet spinning, nozzle-free centrifugal spinning, rotary jet spinning, melt blow spinning and flash-spinning [8,9]. Out of these, electrospinning has been extensively used for the fabrication of fibres; however, it has some limitations. Firstly, it can only be used for polymer systems that are electrically conductive, and secondly, formation of a high fraction of β-phase (which has the highest piezoelectric response) is dependent on very high electric field making the process a safety hazard [10]. As there is electric field involved, it also requires the use of conductive collectors. It also has low yield making it a laborious process and unfit for scale-up demands.

SBS has emerged as an alternative technique to produce sub-micron/nano sized fibres and can relieve the user of the limitations posed by electrospinning. In SBS, a polymer is dissolved into a suitable solvent to reduce its viscosity as thin fibres cannot be produced with very viscous polymer melt [11,12]. The solution is then injected through a nozzle which is surrounded by a concentric outer pipe from which air is purged out. The solution interacts with the air and forms short fibres which fall on a collector. The advantage of SBS is that it can be applied to both electrically conducting and insulating systems and does not require the application of electric field and conductive collectors to initiate fibre processing [13,14]. Moreover, the yield of fibre production is about hundred times higher than that of electrospinning making it suitable for industrialisation [15,16]. The nozzle design is very critical in SBS as it significantly affects the airflow field distribution, the air velocity and morphology of the final product [17]. If the internal diameter of nozzle is too large, outsized droplets will be produced resulting in fibres with larger diameters. Similarly, a very small orifice will reduce the throughput, however, it has the potential of producing thin fibres.

The influence of the protrusion length of the polymer solution syringe on fibre dimensions was found to be insignificant [17]. Lou et al. [18] also reported that the effect of protrusion length has insignificant effect on the fibre morphology. They used four different protrusion lengths: 4, 2, 0 and −2 mm (minus sign shows that the syringe was retracted from the nozzle end by a distance of 2 mm). They reported that the air velocity reaches a maximum in the vicinity of 10–20 mm below the nozzle face. The maximum air velocities were in the range of 170–180 m/s. However, based on practical experiments, they reported that the retracted nozzles resulted in intermittent process with polymer solution blocking the nozzle end. The protruded syringe could produce fibres without such deficiencies.

The attenuation force in solution blow spinning (SBS) is pressurized air and various computational methods have been employed to numerically investigate the influence of air pressure and velocity on the fibre morphology [19–21]. The laminar flow model is considered as the simplest of all available models while the k-ε turbulence model is one of the most commonly used models in computational fluid dynamics (CFD) to simulate mean flow characteristics for turbulent flow conditions with more rapid convergence [18,22]. The k-ε turbulence model is effective for solving problems involving reverse flow [12,23]. It is a semi-empirical model based on model transport equations for the turbulence kinetic energy (k) and its dissipation rate (ε). Neglecting gravitational effects, the transport equations for the k-ε turbulence model are given below [18,24]:

$$\frac{\partial(\rho k)}{\partial t} + \frac{\partial(\rho k u_i)}{\partial x_i} = \frac{\partial}{\partial x_j}\left[\left(\mu + \frac{\mu_t}{\sigma_k}\right)\frac{\partial k}{\partial x_j}\right] + 2\mu_t\left(\frac{\partial u_i}{\partial x_j} + \frac{\partial u_j}{\partial x_i}\right)\frac{\partial u_i}{\partial x_j} - 2\rho\varepsilon M_t^2 \tag{1}$$

$$\frac{\partial(\rho\varepsilon)}{\partial t} + \frac{\partial(\rho\varepsilon u_i)}{\partial x_i} = \frac{\partial}{\partial x_j}\left[\left(\mu + \frac{\mu_t}{\sigma_\varepsilon}\right)\frac{\partial\varepsilon}{\partial x_j}\right] + 2C_{\varepsilon 1}\frac{\varepsilon}{k}\mu_t\left(\frac{\partial u_i}{\partial x_j} + \frac{\partial u_j}{\partial x_i}\right)\frac{\partial u_i}{\partial x_j} - C_{\varepsilon 2}\rho\varepsilon\left(\frac{\varepsilon}{k} + 1\right) \tag{2}$$

where ρ is density kg/m^3, k is turbulent kinetic energy m^2/s^2, t is time s, u_i and u_j are velocity fluctuations in the ith and jth directions, respectively, μ is viscosity kg/(m.s), μ_t is turbulent viscosity kg/(m.s), σ_k and σ_ε are turbulent Prandtl numbers for the kinetic energy and the dissipation rate, respectively, ε is dissipation rate of turbulent kinetic energy, M_t is turbulent Mach number, $C_{\varepsilon 1}$ and $C_{\varepsilon 2}$ are parameters for k-ε turbulence model. The flow characteristics for solution blow spinning process have yet not been studied in detail. For example, when air is passed through the air inlet and moves towards the nozzle tip, whether the nozzle will get choked or not and what will be the influence of choking on the fibre morphology have yet not been reported. A nozzle is choked when the maximum mass flow rate has been reached [25]. Any additional increment in pressure will result in an increase in chamber pressure. Internally the pressure might increase to a value in excess of the rated mechanical strength of the nozzle material which will result in catastrophic failure of the device. Externally of the nozzle, an increase beyond choked conditions can lead to shock wave formation in the nozzle wake. If fluid coming out of the nozzle cannot expand isentropically due to choking an irreversible discontinuity arises called shockwave [26,27]. The shockwave is an abrupt disturbance that causes discontinuous and irreversible changes in fluid characteristics such as speed, density, temperature, and pressure. As a result of the gradient in temperature and velocity being caused by the shock, heat is transferred, and energy is dissipated within the gas. These processes are thermodynamically irreversible [28]. As the nozzle design and the attenuation force (pressurized air) are of utmost importance in SBS, both numerical (CFD) and experimental methods were used to investigate the fibre formation. CFD was used to investigate the convergence point for high speed air as it comes out of the nozzle. The polymer solution syringe was positioned such that it did not choke due to the reversal of air flow. The produced fibres did not show any bead formation at higher pressure values indicating that optimized SBS can successfully produce submicron PVDF fibres.

2. Experimental Work

2.1. Materials

PVDF (Kynar, melt viscosity: 23.0–29.0 MPa.s at room temperature) was supplied by ARKEMA (King of Prussia, PA, USA). N, N-Dimethyl Formamide (anhydrous, 98%) was purchased from Loba Chemie, Mumbai, India. Chemicals were used as received.

2.2. Nozzle Design

A bespoke SBS concentric nozzle was used to produce PVDF submicron fibres and was supplied by AREKA group, Istanbul, Turkey. A schematic of the nozzle used for the CFD model to predict air flow characteristics is shown in Figure 1. The polymer solution syringe is inserted through the opening at the left end of the nozzle that passes all the way through and comes out from the right end of the nozzle. In the middle is an air inlet (~4 mm diameter) that transfers air into the air chamber. To build pressure, the air chamber has four holes with internal diameters of ~1 mm each. These holes are covered with a cap that leaves a very narrow fissure for the air to come out 360° around the concentric polymer solution syringe. To ease manufacturing, the nozzle consists of four metallic sections (Figure 2) which are assembled to form the nozzle unit. Consequently, the experiments were carried out using this assembled nozzle. Most of the CFD models in the reviewed literature were carried out in 2D systems which were then extrapolated to 3D [18]. However, in this study, a 3D nozzle was employed which was also used in the experiments and using the same system for theoretical calculations will provide a more realistic comparison.

2.3. Experimental Setup

The experimental setup is shown in Figure 3 and a schematic of the process is shown in Figure 4. As variation in temperature can influence fluid viscosity, and rheological properties are dependent on viscosity, it was ensured that the temperature of the polymer solution was held constant by keeping the SBS setup in a ventilated fume hood. No heating or cooling mechanism was used in the fume hood and the temperature during experiments remained in the range 22–25 °C. Pressurized air was used to attenuate polymer solution jet at 1–5 bar generated by Cruiser air compressor (1.5 HP, 30 L, Shanghai, China). NE-300 infusion syringe pump with 21-gauge needle was employed to feed the polymer solution, and was positioned inside the concentric nozzle with internal diameter d_i = 2 mm. A working distance of 20 cm was maintained between the nozzle and the drum collector.

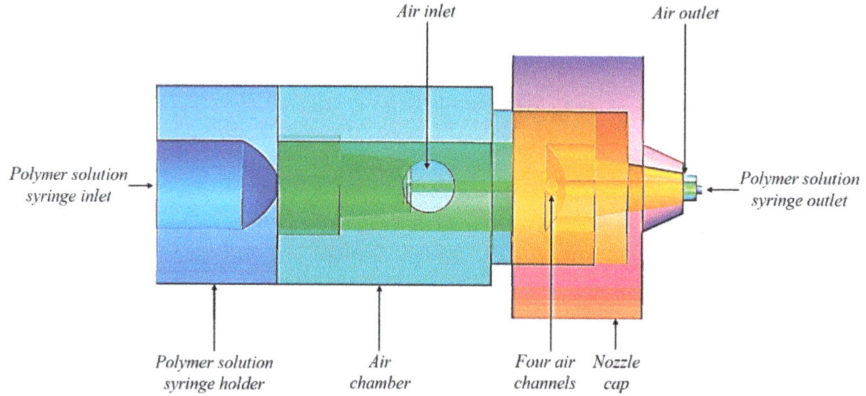

Figure 1. Different sections of solution blow spinning (SBS) nozzle.

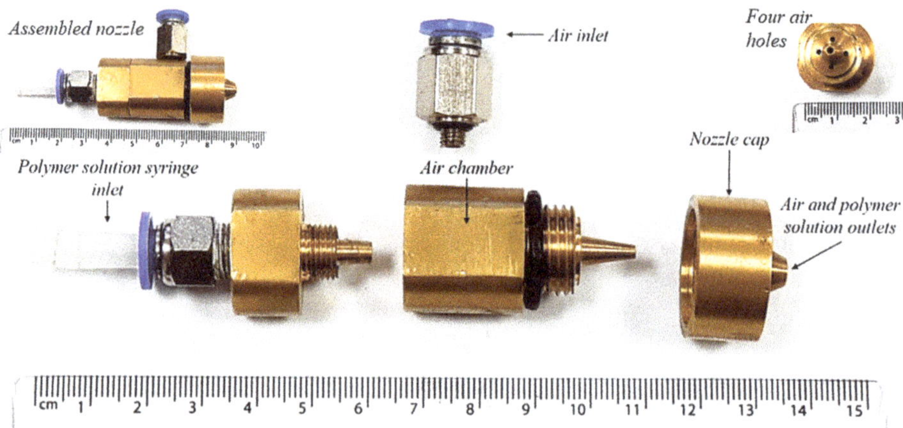

Figure 2. Various parts of the fabricated nozzle with four air holes through air chamber shown in the top right corner and assembled nozzle shown in the top left corner.

Figure 3. SBS setup for the production of polyvinylidene fluoride (PVDF) fibres.

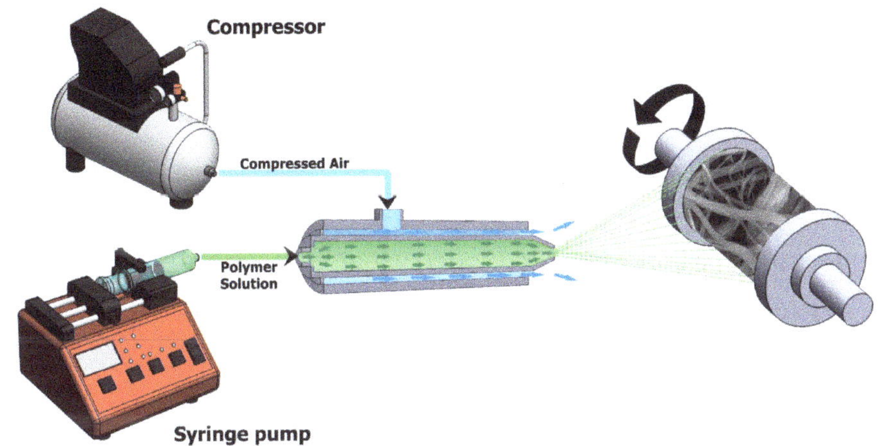

Figure 4. Schematic of solution blow spinning process.

2.4. Production of Samples/Fibres

A quantity of 15 wt % of PVDF polymer solution was pumped with the aid of the syringe and the feed rate of polymer solution was varied between 2 to 10 mL/h. Air pressure was varied between 1 to 5 bar (only one variable was changed at a time). The fibres were collected at the drum collector and analysed for their morphological features.

2.5. Characterization

CFD was carried out in ANSYS Fluent v19 (ANSYS Inc., Canonsburg, PA, USA). For the CFD simulation, a solid enclosure was built around the nozzle with zero shear slip and meshed consisting of 793,731 nodes and 3,355,960 elements. A steady-state, compressible Navier–Stokes equations was used whose component forms for continuity, x-momentum, y-momentum, z-momentum, and energy are given below, respectively, where p is pressure, q is heat flux, u, v, and w are velocity components, E_T is total energy, Re is Reynolds number, and Pr is Prandtl number.

$$\frac{\partial(\rho u)}{\partial x} + \frac{\partial(\rho v)}{\partial y} + \frac{\partial(\rho w)}{\partial z} = 0 \tag{3}$$

$$\frac{\partial(\rho u^2)}{\partial x} + \frac{\partial(\rho uv)}{\partial y} + \frac{\partial(\rho uw)}{\partial z} = -\frac{\partial p}{\partial x} + \frac{1}{Re}\left[\frac{\partial \tau_{xx}}{\partial x} + \frac{\partial \tau_{xy}}{\partial y} + \frac{\partial \tau_{xz}}{\partial z}\right] \tag{4}$$

$$\frac{\partial(\rho uv)}{\partial x} + \frac{\partial(\rho v^2)}{\partial y} + \frac{\partial(\rho vw)}{\partial z} = -\frac{\partial p}{\partial y} + \frac{1}{Re}\left[\frac{\partial \tau_{xy}}{\partial x} + \frac{\partial \tau_{yy}}{\partial y} + \frac{\partial \tau_{yz}}{\partial z}\right] \tag{5}$$

$$\frac{\partial(\rho uw)}{\partial x} + \frac{\partial(\rho vw)}{\partial y} + \frac{\partial(\rho w^2)}{\partial z} = -\frac{\partial p}{\partial z} + \frac{1}{Re}\left[\frac{\partial \tau_{xz}}{\partial x} + \frac{\partial \tau_{yz}}{\partial y} + \frac{\partial \tau_{zz}}{\partial z}\right] \tag{6}$$

$$\begin{aligned}
\frac{\partial(uE_T)}{\partial x} + \frac{\partial(vE_T)}{\partial y} &+ \frac{\partial(wE_T)}{\partial z} \\
&= -\frac{\partial(up)}{\partial x} - \frac{\partial(vp)}{\partial y} - \frac{\partial(wp)}{\partial z} - \frac{1}{RePr}\left[\frac{\partial q_x}{\partial x} + \frac{\partial q_y}{\partial y} + \frac{\partial q_z}{\partial z}\right] \\
&+ \frac{1}{Re}\left[\frac{\partial}{\partial x}\left(u\tau_{xx} + v\tau_{xy} + w\tau_{xz}\right) + \frac{\partial}{\partial y}\left(u\tau_{xy} + v\tau_{yy} + w\tau_{yz}\right)\right. \\
&\left. + \frac{\partial}{\partial z}\left(u\tau_{xz} + v\tau_{yz} + w\tau_{zz}\right)\right]
\end{aligned} \tag{7}$$

The closure models used were the ideal gas model for density and the Sutherland equation for viscosity. The constants used were for air and mentioned below:

$$\mu = \mu_0 \left(\frac{T_0 + C}{T + C} \right) \left(\frac{T}{T_0} \right)^{3/2}$$

(8)

where, T_0 is reference temperature (K), μ_0 is reference viscosity (1.716×10^{-5} kg/m.s) at the reference temperature T_0 (273 K), T is effective temperature (110 K), and C is Sutherland's constant for given gaseous material. The outlets were all specified as pressure far-field outlets. The inlet to the nozzle was a pressure inlet and remainder were all no-slip wall boundaries. Both laminar and k-ε turbulent simulations were carried out. Although the flow field is inherently turbulent, laminar flow was investigated as a steppingstone and was used to compare the results with turbulent flow simulation. The laminar results are not shown here but can be viewed in the Supplementary Materials (Figures S1–S22). A scanning electron microscope (JSM-6010LV-SEM, JEOL, Tokyo, Japan) was used to investigate the morphology of the produced fibres. Fibre diameter was analysed using Image-J software, average fibre diameter distribution was detected by measuring the distance across fibre boundaries at different imaging scales to ensure consistency.

3. Results and Discussion

Under both laminar and k-ε turbulent flow conditions, a reverse flow was observed in the vicinity of polymer solution syringe outlet as shown in Figure 5. If a polymer solution droplet gets trapped by this reverse flow, there is high possibility that it will not move forward and will choke the syringe/nozzle. This supposition is further supported by the experimental observations reported by Lou et al. [18] who found that the retracted nozzle resulted in an intermittent process with polymer solution blocking the nozzle end. They also reported that the best morphology of fibres was produced when polymer syringe was protruded out by 4 mm.

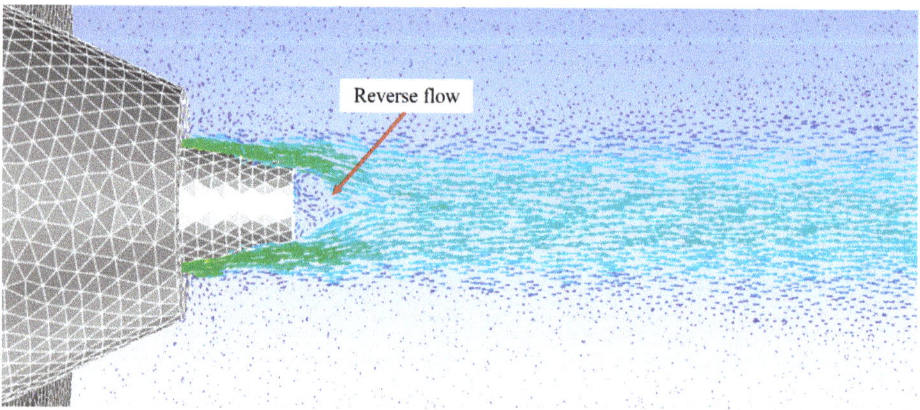

Figure 5. Reverse flow in the vicinity of the polymer solution syringe tip that can choke the syringe.

Velocity, pressure, temperature, turbulent kinetic energy, and density contours at different air pressures are shown in Figures 6 and 7. A magnified set of contours at 4 bar is shown in Figure 8 as the thinnest fibres were produced at that pressure. For clarity of readings large contours have been provided in the Supplementary Materials (Figures S23–S71). Velocity, temperature, and turbulent kinetic energy profiles at six vertical slices (at 0, 1, 4, 6, 16, and 26 mm from the nozzle end) were also calculated and plotted in Figures 9–11, respectively. The values of the parameters and z-velocity were also measured at the horizontal symmetry axis of the geometry and are shown in Figures 12 and 13.

They will be referred to as centreline air velocity (v_c), pressure (P_c), temperature (T_c), turbulent kinetic energy (TKE_c), and density (ρ_c) from hereafter.

At 1 bar, v_c increases quadratically to ~80 m/s at a distance of ~0.7 mm from the air outlet, then slows down due to reverse flow until hits zero, and then shoots up to ~120 m/s at a distance of ~7 mm and then starts to gradually decrease. The exact value of air velocity may change due to polymer solution droplets in the flow field. Due to flow reversal, the P_c initially decreases quadratically to 3 kPa at a distance of ~0.5 mm, after which it starts to increase until hits zero at a distance of ~1 mm, saturates at ~3 kPa at a distance of ~1.7 mm, and then gradually flattens out. The minimum T_c recorded at ~4 mm was 291 K (18 °C) which was 7 K cooler than the ambient temperature which was set at 298 K (25 °C). The maximum TKE_C was ~1 kJ/kg recorded at ~1.3 mm from the air outlet. Experiments were conducted at 1 bar in order to produce fibres using the setup shown in Figure 3. Despite changing the feed rate between 2 mL/h to 10 mL/h, fibres could not be formed. This was ascribed to that the polymer solution droplets kept falling on the floor just ahead of the nozzle. Therefore, it is believed that air velocity of 120 m/s is not high enough to generate PVDF fibres. Similar trends were observed at 2–10 bars for CFD results; however, fibre morphology was significantly affected by air pressure.

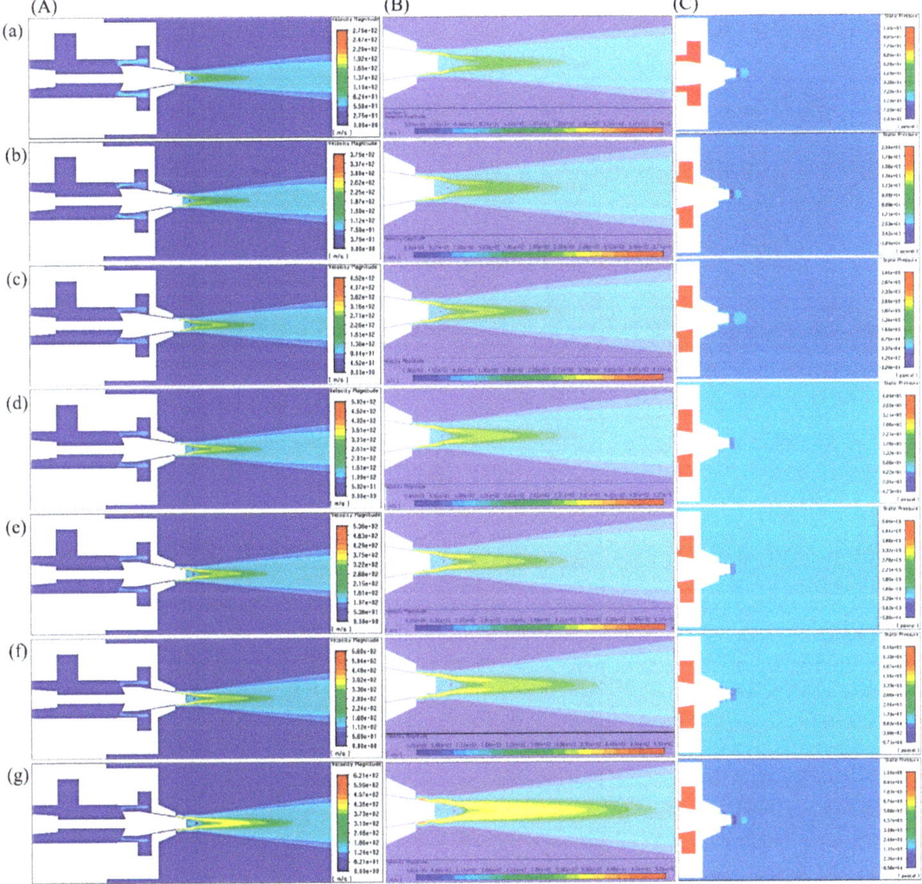

Figure 6. k-ε turbulence flow contours through SBS nozzle at different air pressures: (**A**,**B**) velocity (m/s), (**C**) pressure (Pa), (**a**) 1 bar, (**b**) 2 bar, (**c**) 3 bar, (**d**) 4 bar, (**e**) 5 bar, (**f**) 6 bar, (**g**) 10 bar.

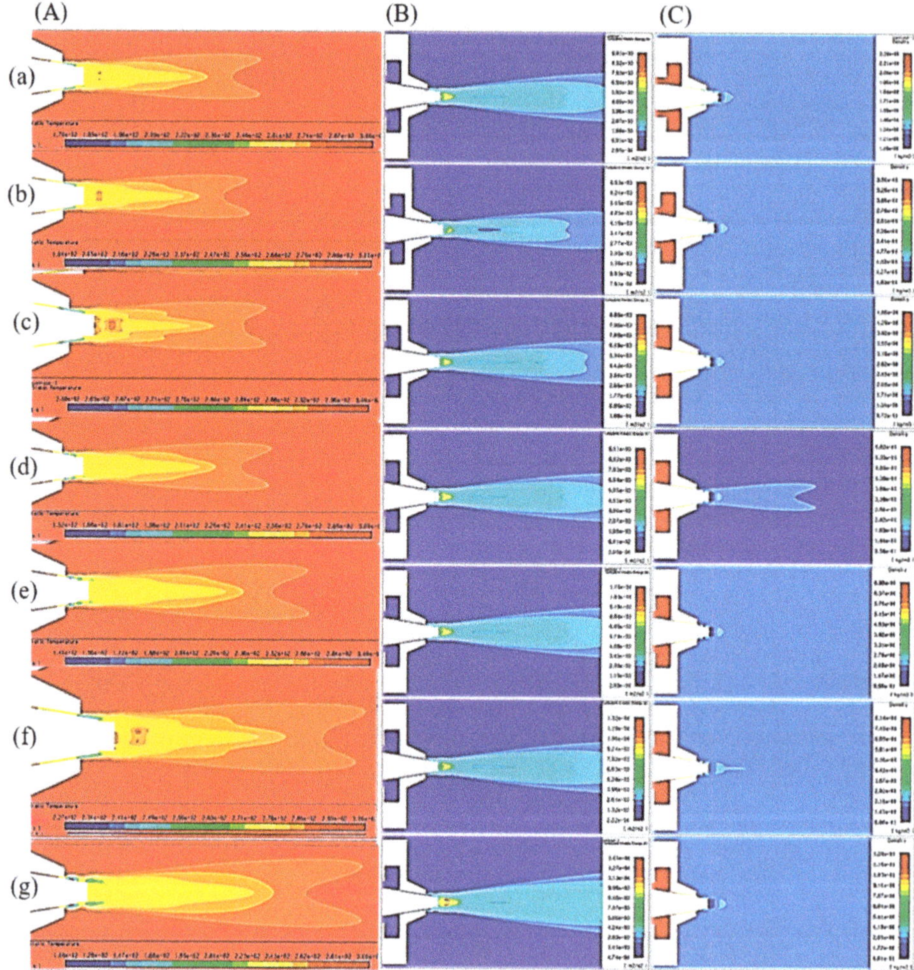

Figure 7. k-ε turbulence flow contours through SBS nozzle at different air pressures: (**A**) temperature (K), (**B**) turbulent kinetic energy (J/kg = m²/s²), and (**C**) density (kg/m³), (**a**) 1 bar, (**b**) 2 bar, (**c**) 3 bar, (**d**) 4 bar, (**e**) 5 bar, (**f**) 6 bar, (**g**) 10 bar.

Fibres were successfully produced at 2 bar whose SEM image is shown in Figure 14a and diameter distribution in Figure 14b. Fibres with wide distribution of diameters (between 100–900 nm) were observed and it is important to mention that some beads were also observed on these fibres. At 3 bar, relatively thinner fibres (than at 2 bar) as shown in Figure 14c were produced. The fibres were almost defect free with fibre diameters in the range of 140–700 nm (Figure 14d). At 4 bar, Much thin and bead-free fibres were produced as shown in Figure 14e. Most of the fibres were in 100–350 nm range (Figure 14f) and these were the thinnest fibres produced in the current work. The fibres produced at 5 bar were relatively thick as shown in Figure 14g. There were some intertwined fibre ropes observed at 5 bar and diameter of each rope was measured to get mean fibre diameter of ~1.5 μm (Figure 14h). At 5 bar, there is drop in temperature due to Joule-Thomson effect up to ~251 K (−22 °C). This cryogenic environment can produce residual stresses in the fibres that might have caused them turn and twist around each other while either during flight or after hitting the drum collector and resulted in interlocked fibre ropes. The turbulent kinetic energy (Figure 13c,d) continues to rise with increasing air pressure and velocity and this increased turbulence might be another factor in destabilizing the fibres in flight.

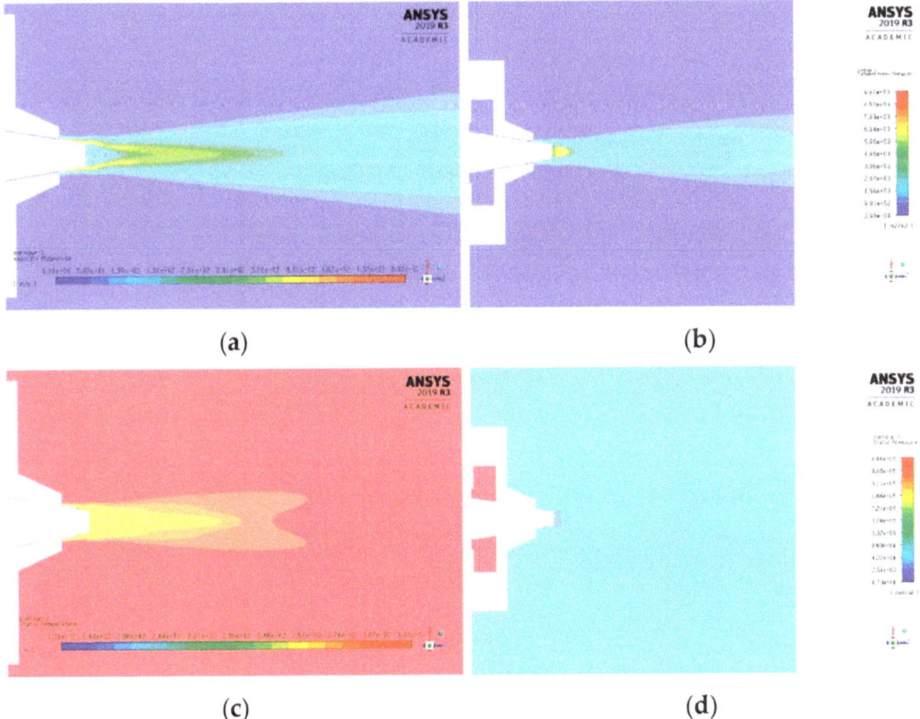

Figure 8. k-ε turbulence flow contours through SBS nozzle at air pressure of 4 bar: (**a**) velocity (m/s), (**b**) turbulent kinetic energy (J/kg = m^2/s^2), (**c**) temperature (K), and (**d**) pressure (Pa).

The z-velocity graphs (Figure 12c,d) hit negative values that confirm reverse flow. It is important to mention that the reverse flow is at the tip of syringe, which suggests that the polymer solution needs to be injected forcibly to pass the barrier created by the reverse flow (Figure 12e,f). There is fluctuation in the temperature as well (Figure 13a,b) due to reverse flow. As air leaves the air outlet, the temperature follows a sinusoidal curve dropping at a distance of ~1 mm from the air outlet, which goes up due to reverse flow, then drops again until it hits the lowest value (~193 K (−80 °C) at 10 bar) and finally gradually goes up to the room temperature. A similar trend was observed for turbulent kinetic energy (Figure 13c,d) along the positive ordinate.

The density plots (Figure 13e,f) show that the air is less dense at the nozzle opening. Air imposes frictional drag on the moving objects which is approximately proportional to the square of the velocity of the moving object. A decrease in density and viscosity would mean less frictional drag being exerted on the polymer solution droplet through the air flow field which would be able to move faster and will get thinner under the influence of attenuating force, i.e., high-speed air. The density graphs hit the peak values at ~4 mm from the nozzle end. This increment in density is propitious as it would exert compressive stresses on the fibres thereby resulting in compact fibres.

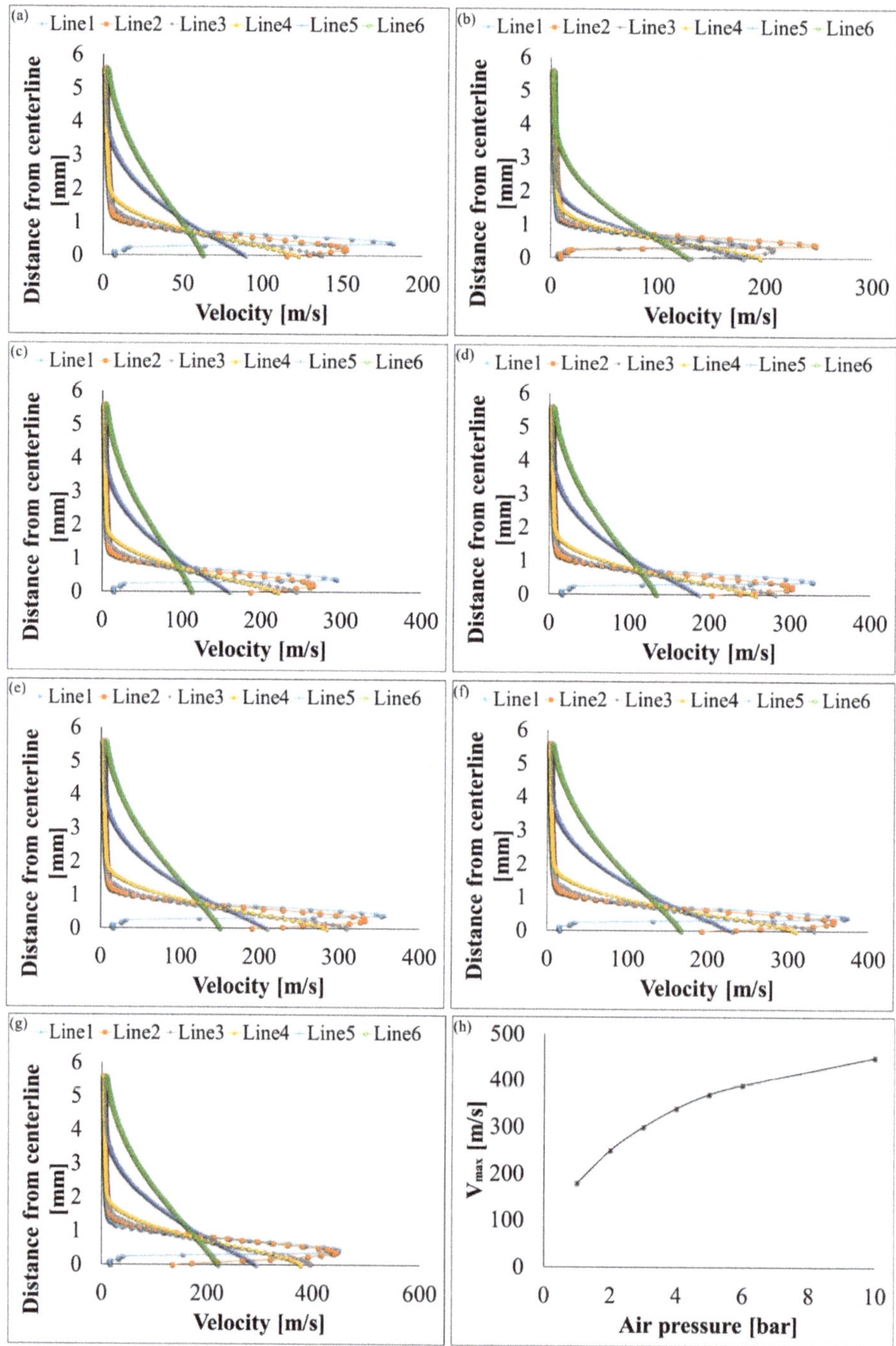

Figure 9. Velocity profiles at six vertical slices (at distance of 0, 1, 4, 6, 16, and 26 mm from the air outlet) for k-ε turbulence flow plots through SBS nozzle at different air pressures: (**a**) 1 bar, (**b**) 2 bar, (**c**) 3 bar, (**d**) 4 bar, (**e**) 5 bar, (**f**) 6 bar, (**g**) 10 bar, and (**h**) maximum velocity values at different air pressures.

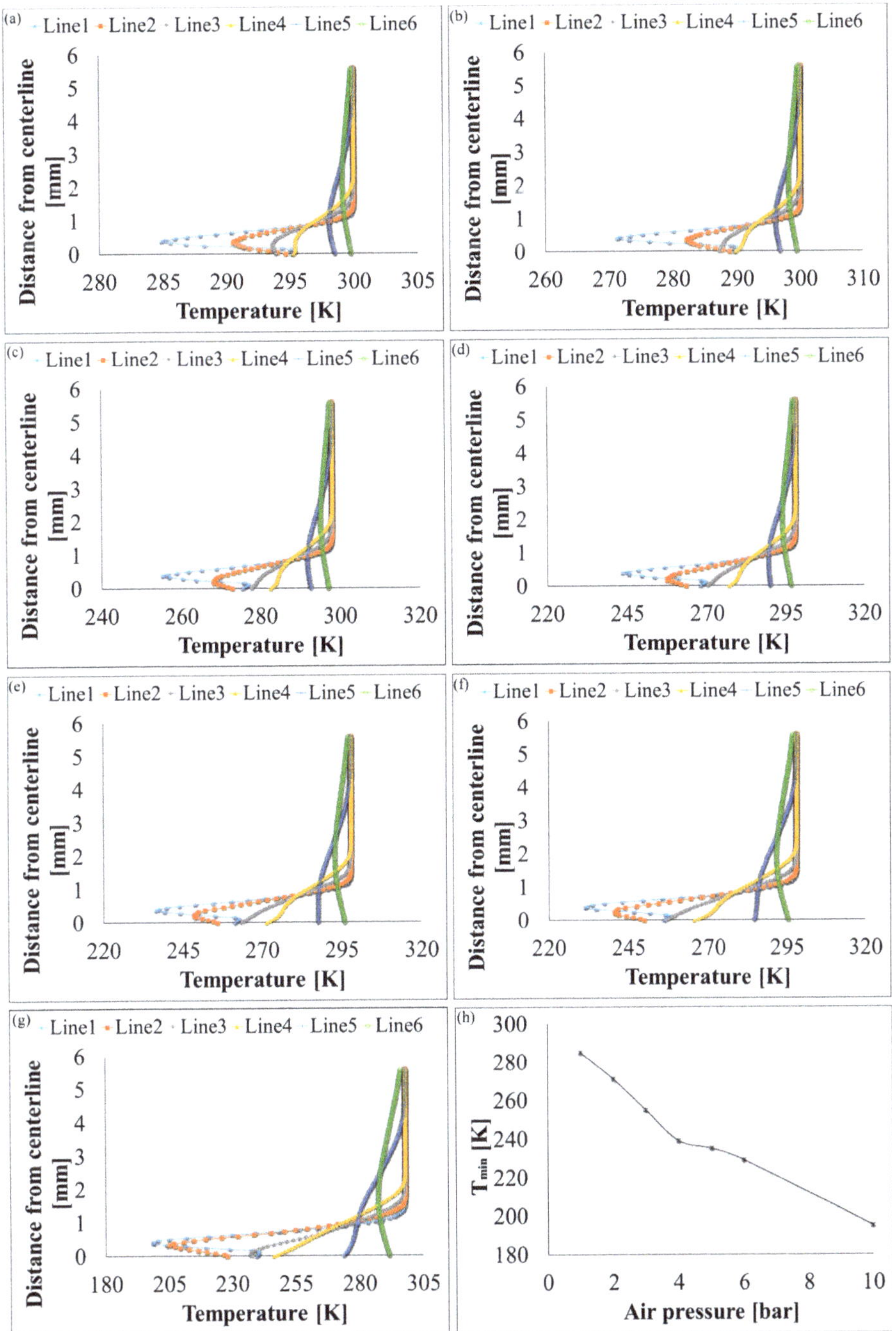

Figure 10. Temperature profiles at six vertical slices (at distance of 0, 1, 4, 6, 16, and 26 mm from the air outlet) for k-ε turbulence flow plots through SBS nozzle at different air pressures: (**a**) 1 bar, (**b**) 2 bar, (**c**) 3 bar, (**d**) 4 bar, (**e**) 5 bar, (**f**) 6 bar, (**g**) 10 bar, and (**h**) minimum temperature values at different air pressures.

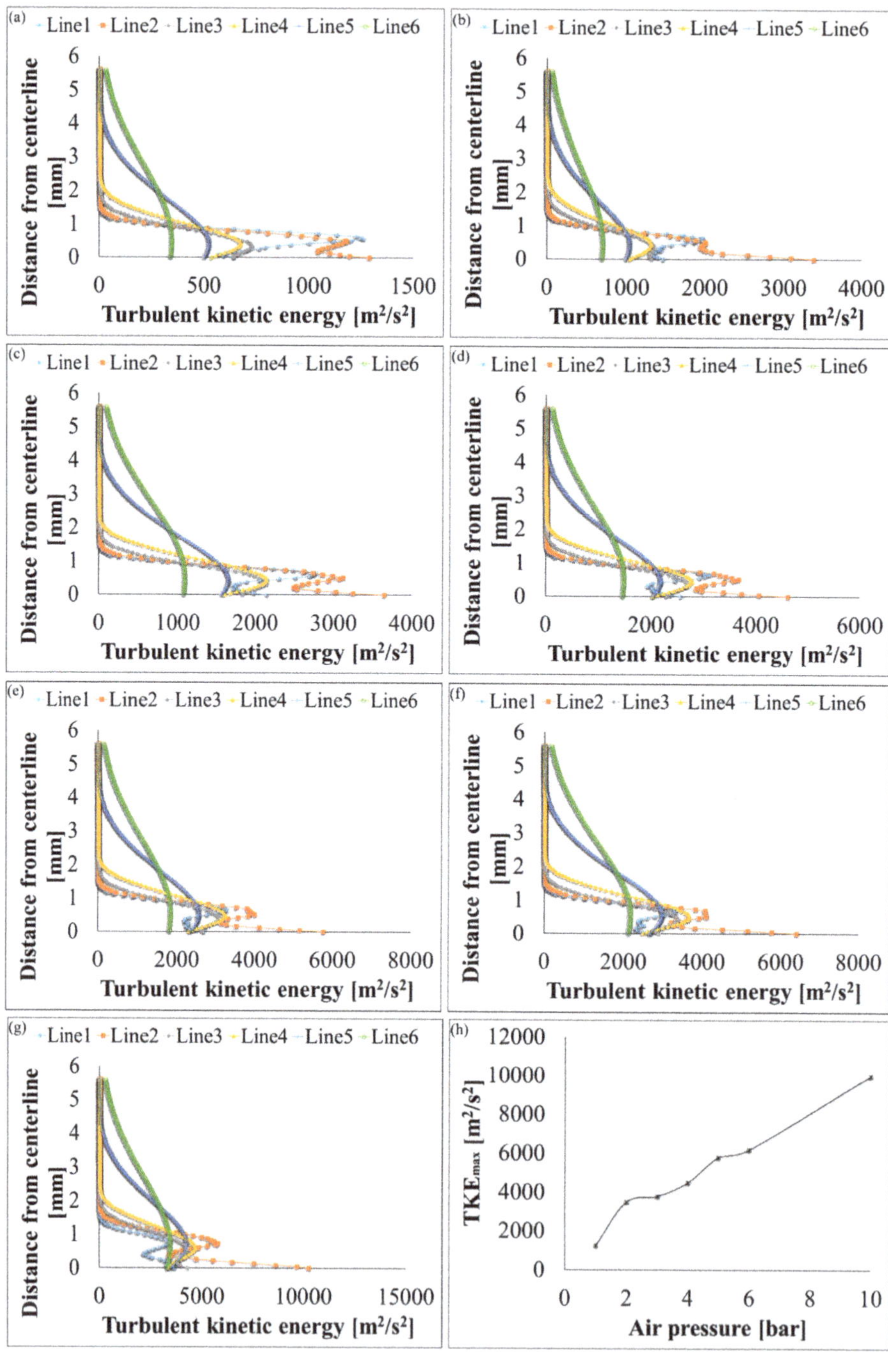

Figure 11. Turbulent kinetic energy profiles at six vertical slices (at distance of 0, 1, 4, 6, 16, and 26 mm from the air outlet) for k-ε turbulence flow plots through SBS nozzle at different air pressures: (**a**) 1 bar, (**b**) 2 bar, (**c**) 3 bar, (**d**) 4 bar, (**e**) 5 bar, (**f**) 6 bar, (**g**) 10 bar, and (**h**) maximum turbulent kinetic energy values at different air pressures.

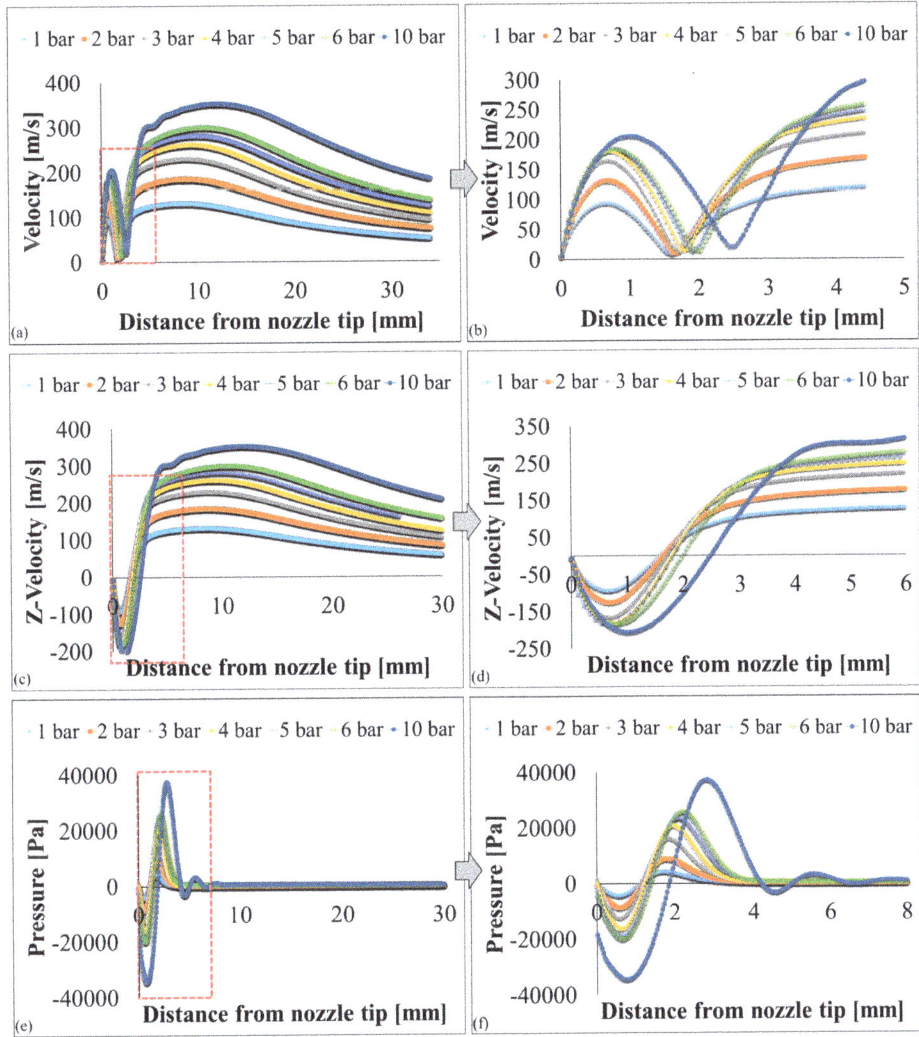

Figure 12. k-ε turbulence flow plots through SBS nozzle at different air pressures: (**a**,**b**) velocity (m/s), (**c**,**d**) z-velocity (m/s), and (**e**,**f**) pressure (Pa).

A significant drop in temperature was observed as air comes out of the nozzle. It can possibly be related to the Joule–Thomson effect inducing temperature drops when high speed fluid quickly escapes through a narrow hole [29]. Static temperature drops during the isentropic expansion process and this drop in temperature is even more evident under supersonic flow where temperature can drop down to 213 K (−60 °C) without any cryogenic cooling or use of solid adsorption techniques [28]. It has been reported that PVDF solution temperature influences spinnability of PVDF fibres [15]. The drop in temperature can alter viscosity of the polymer solution as viscosity and temperature are inversely related to each other. When temperature of polymer solution is high, its viscosity will be low, and therefore low attenuation force will suffice to get thin fibres and vice versa. Attenuation force in SBS is high speed air which means that thin fibres can be achieved at relatively low air pressure and velocity. A higher air pressure results in a higher air velocity and turbulent fluctuations. The higher air velocity will result in thinner fibres but turbulent fluctuations may break the fibres [18]. It has been reported that there is a direct relationship between viscosity of polymer solution and mean fibre diameter. Haddadi et al. [30] incorporated hydrophobic and hydrophilic nanosilica into PVDF and reported that mean fibre diameter increased in both cases. They suggested that the viscosity of polymer solution increased by the incorporation of nanofillers which in turn led to an increase in mean fibre diameter. Yun et al. [31] fabricated $Pb(Zr_{0.53}Ti_{0.47})O_3$ reinforced PVDF nanofibres and reported that both density and viscosity of the polymer solution increased after the incorporation of PZT. The mean fibre diameter increased until 10 wt % and then gradually decreased when volume fraction was further increased up to 30 wt % [31].

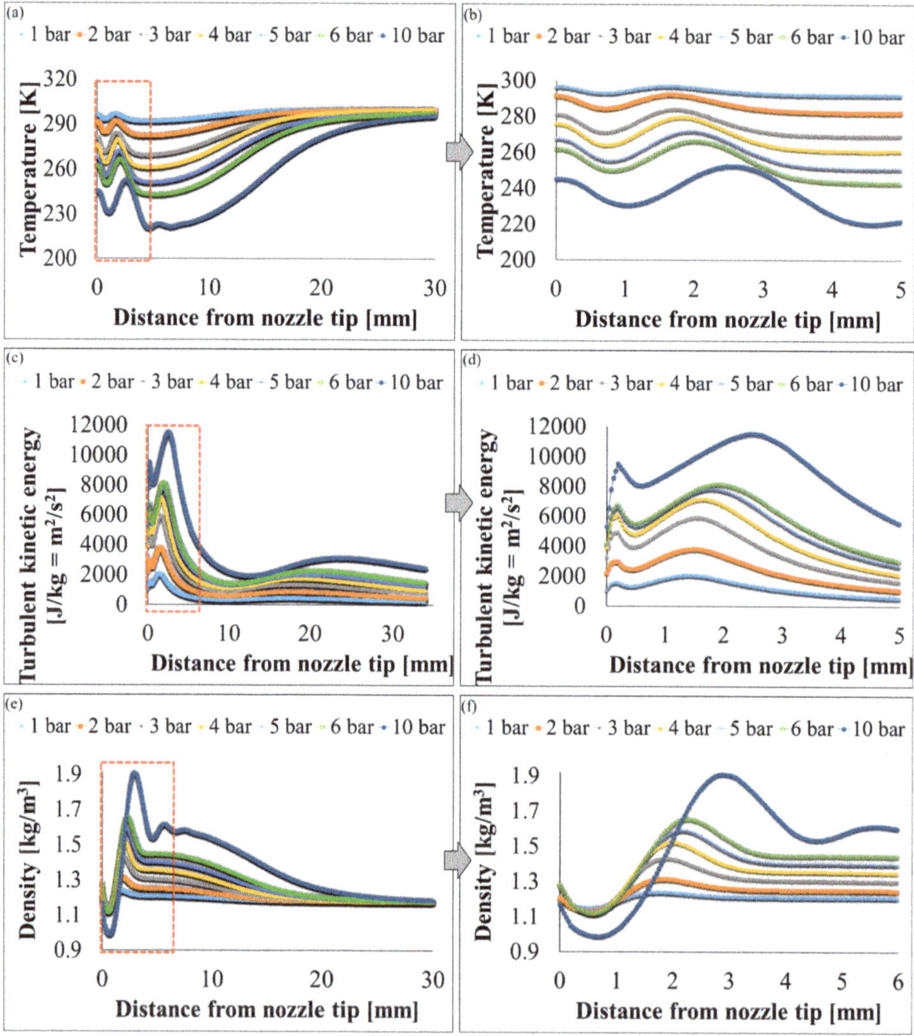

Figure 13. k-ε turbulence flow plots through SBS nozzle at different air pressures: (**a,b**) temperature (K), (**c,d**), turbulent kinetic energy (J/kg = m^2/s^2), and (**e,f**) density [kg/m^3].

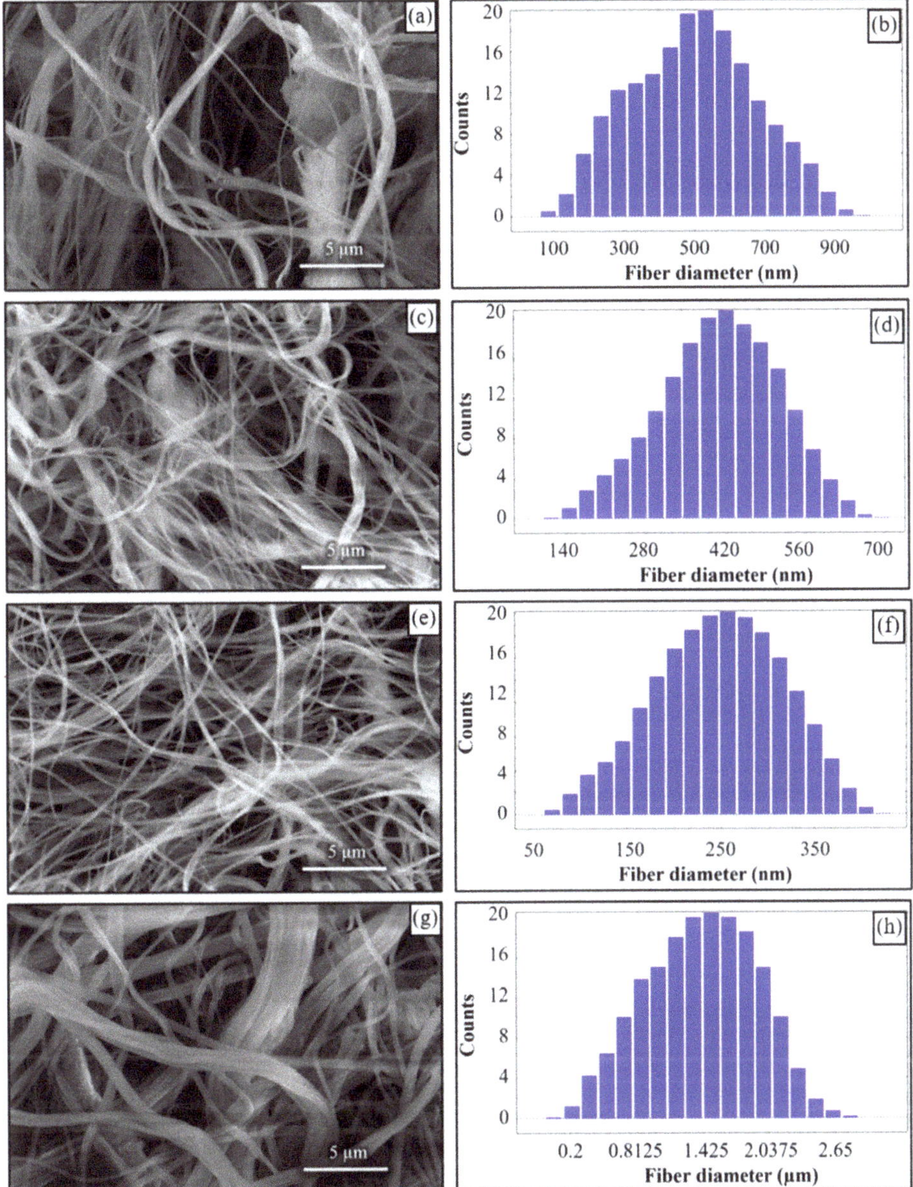

Figure 14. (**a,c,e,g**) SEM results at 2,3,4, and 5 bar air pressures, and (**b,d,f,h**) their respective fibre size distribution.

The v_c increases at 6 bar and 10 bar to ~400 and ~450 m/s, respectively. Experiments at 10 bar could not be performed due to the limitation of the setup. However, CFD results provide a useful information that the nozzle did not choke up until 10 bar. The effect of shock structures on the fibre formation has not been determined, however it is likely that the rapidly changing conditions before and after the shock will have a detrimental effect on the fibre morphology [32]. To avoid choking, nozzle diameter, feed rate and air pressure must be carefully optimized [32,33].

4. Conclusions

The current research provides a comprehensive numerical and experimental analysis for fibres produced using solution blow spinning (SBS). The flow characteristics of high-speed air through a bespoke nozzle were investigated numerically using computational fluid dynamics (CFD) and experimentally in a custom-built SBS setup to produce polyvinylidene fluoride (PVDF) submicron fibres. It has been concluded that under both laminar and k-ε turbulent flow conditions, a reverse flow was observed in the vicinity of polymer solution syringe outlet. As a result of this, if polymer solution droplets got trapped by this reverse flow, there is high possibility that the droplets will not move forward causing the syringe/nozzle to choke. However, chocking of nozzle does not take place at air pressures up to 10 bar. Submicron fibres were produced at air pressures ≥ 2 bar. However, higher pressure of 4 bar was found to be more effective in producing fibres in the range of 150–200 nm. Upon increasing the air pressure to 5 bar, the fibre diameter increased and interlaced fibre ropes were produced as significant temperature drop (~251 K (−22 °C)) was observed due to Joule–Thomson effect. The results presented in the paper will pave the way for future research in fibre manufacturing using SBS.

Supplementary Materials: The following are available online at http://www.mdpi.com/2073-4360/12/5/1140/s1, Figure S1. Solid enclosure with zero slip around SBS nozzle. Figure S2. Meshed system for CFD of SBS nozzle. The mesh comprises of 793,731 nodes with 33,55,960 elements. Figures S3–S12. Laminar flow results. Figures S13–S22. CFD contours for k-ε turbulent model at different values of ε_{c1} and ε_{c2}. Figures S23–S71. k-ε turbulent model-based contours with default parameters.

Author Contributions: I.S. and A.H.H. conceived the project; E.E., N.S. and A.H.H. carried out the experimental work; M.C. and R.A. carried out the CFD; I.S. supervised all project activities; R.A. analysed SEM images and wrote the manuscript; J.K. supported manuscript writing and editing. All authors read and refined the data to make a compelling and coherent writing. All authors have read and agreed to the published version of the manuscript.

Funding: This research was funded by British Council grant number 352360451 and Newton-Mosharafa Call between UK and Egypt, ID: 30886.

Acknowledgments: The authors would like to thank British Council for providing funding for the Institutional Links Project (Project ID: 352360451) between Alexandria University, Egypt, and Northumbria University, Newcastle upon Tyne, UK. This work is part of the project of (Newton-Mosharafa Call between UK and Egypt, ID: 30886) which has been funded by Science, Technology & Innovation Funding Authority (STIFA), Egypt.

References

1. Kawai, H. The Piezoelectricity of Poly (vinylidene Fluoride). *Jpn. J. Appl. Phys.* **1969**, *8*, 975–976. [CrossRef]
2. Persano, L.; Dagdeviren, C.; Su, Y.; Zhang, Y.; Girardo, S.; Pisignano, D.; Huang, Y.; Rogers, J.A. High performance piezoelectric devices based on aligned arrays of nanofibers of poly(vinylidenefluoride-co-trifluoroethylene). *Nat. Commun.* **2013**, *4*, 1610–1633. [CrossRef] [PubMed]
3. Wang, Z.; Tan, L.; Pan, X.; Liu, G.; He, Y.; Jin, W.; Li, M.; Hu, Y.; Gu, H. Self-Powered Viscosity and Pressure Sensing in Microfluidic Systems Based on the Piezoelectric Energy Harvesting of Flowing Droplets. *ACS Appl. Mater. Interfaces* **2017**, *9*, 28586–28595. [CrossRef] [PubMed]
4. Lee, H.S.; Chung, J.; Hwang, G.-T.; Jeong, C.K.; Jung, Y.; Kwak, J.-H.; Kang, H.; Byun, M.; Kim, W.D.; Hur, S.; et al. Flexible inorganic piezoelectric acoustic nanosensors for biomimetic artificial hair cells. *Adv. Funct. Mater.* **2014**, *24*, 6914–6921. [CrossRef]
5. Seo, M.H.; Yoo, J.Y.; Choi, S.Y.; Lee, J.S.; Choi, K.; Jeong, C.K.; Lee, K.J.; Yoon, J.B. Versatile Transfer of an Ultralong and Seamless Nanowire Array Crystallized at High Temperature for Use in High-Performance Flexible Devices. *ACS Nano* **2017**, *11*, 1520–1529. [CrossRef]
6. Zou, H.; Li, X.; Peng, W.; Wu, W.; Yu, R.; Wu, C.; Ding, W.; Hu, F.; Liu, R.; Zi, Y.; et al. Piezo-Phototronic Effect on Selective Electron or Hole Transport through Depletion Region of Vis–NIR Broadband Photodiode. *Adv. Mater.* **2017**, *29*, 1–10. [CrossRef]
7. Guo, R.; Guo, Y.; Duan, H.; Li, H.; Liu, H. Synthesis of Orthorhombic Perovskite-Type ZnSnO3 Single-Crystal Nanoplates and Their Application in Energy Harvesting. *ACS Appl. Mater. Interfaces* **2017**, *9*, 8271–8279. [CrossRef]

8. Zhang, Y.; Huang, Z.M.; Xu, X.; Lim, C.T.; Ramakrishna, S. Preparation of core-shell structured PCL-r-gelatin bi-component nanofibers by coaxial electrospinning. *Chem. Mater.* **2004**, *16*, 3406–3409. [CrossRef]

9. Xu, X.; Zhuang, X.; Chen, X.; Wang, X.; Yang, L.; Jing, X. Preparation of core-sheath composite nanofibers by emulsion electrospinning. *Macromol. Rapid Commun.* **2006**, *27*, 1637–1642. [CrossRef]

10. Jiyong, H.; Yuanyuan, G.; Hele, Z.; Yinda, Z.; Xudong, Y. Effect of electrospinning parameters on piezoelectric properties of electrospun PVDF nanofibrous mats under cyclic compression. *J. Text. Inst.* **2017**, *109*, 843–850. [CrossRef]

11. Wang, Y.; Wang, X. Investigation on a new annular melt-blowing die using numerical simulation. *Ind. Eng. Chem. Res.* **2013**, *52*, 4597–4605. [CrossRef]

12. Sun, Y.F.; Liu, B.W.; Wang, X.H.; Zeng, Y.C. Air-Flow Field of the Melt-Blowing Slot Die via Numerical Simulation and Multiobjective Genetic Algorithms. *J. Appl. Polym. Sci.* **2011**, *122*, 3520–3527. [CrossRef]

13. Vural, M.; Behrens, A.M.; Ayyub, O.B.; Ayoub, J.J.; Kofinas, P. Sprayable elastic conductors based on block copolymer silver nanoparticle composites. *ACS Nano* **2014**, *9*, 336–344. [CrossRef] [PubMed]

14. Zhang, L.; Kopperstad, P.; West, M.; Hedin, N.; Fong, H. Generation of Polymer Ultrafine Fibers Through Solution (Air-) Blowing. *J. Appl. Polym. Sci.* **2009**, *114*, 3479–3486. [CrossRef]

15. Tandon, B.; Kamble, P.; Olsson, R.T.; Blaker, J.J.; Cartmell, S.H. Fabrication and characterisation of stimuli responsive piezoelectric PVDF and hydroxyapatite-filled PVDF fibrous membranes. *Molecules* **2019**, *24*, 1903. [CrossRef]

16. Magaz, A.; Roberts, A.D.; Faraji, S.; Nascimento, T.R.L.; Medeiros, E.S.; Zhang, W.; Greenhalgh, R.; Mautner, A.; Li, X.; Blaker, J.J. Porous, Aligned, and Biomimetic Fibers of Regenerated Silk Fibroin Produced by Solution Blow Spinning. *Biomacromolecules* **2018**, *19*, 4542–4553. [CrossRef]

17. Han, W.; Xie, S.; Sun, X.; Wang, X.; Yan, Z. Optimization of airflow field via solution blowing for chitosan/PEO nanofiber formation. *Fibers Polym.* **2017**, *18*, 1554–1560. [CrossRef]

18. Lou, H.; Han, W.; Wang, X. Numerical study on the solution blowing annular jet and its correlation with fiber morphology. *Ind. Eng. Chem. Res.* **2014**, *53*, 2830–2838. [CrossRef]

19. Drabek, J.; Zatloukal, M. Meltblown technology for production of polymeric microfibers/nanofibers: A review. *Phys. Fluids* **2019**, *31*, 091301. [CrossRef]

20. Hassan, M.A.; Anantharamaiah, N.; Khan, S.A.; Pourdeyhimi, B. Computational Fluid Dynamics Simulations and Experiments of Meltblown Fibrous Media: New Die Designs to Enhance Fiber Attenuation and Filtration Quality. *Ind. Eng. Chem. Res.* **2016**, *55*, 2049–2058. [CrossRef]

21. Moore, E.M.; Shambaugh, R.; Papavassiliou, D.V. Papavassiliou, Analysis of isothermal annular jets: Comparison of computational fluid dynamics and experimental data. *J. Appl. Polym. Sci.* **2004**, *94*, 909–922. [CrossRef]

22. Gopalakrishnan, R.N.; Disimile, P.J. Effect of Turbulence Model in Numerical Simulation of Single Round Jet at Low Reynolds Number. *Int. J. Comput. Eng. Res.* **2017**, *7*, 29–44.

23. Gopalakrishnan, R.N.; Disimile, P.J. CFD analysis of twin turbulent impinging round jets at different impingement angles. *Fluids* **2018**, *3*, 79. [CrossRef]

24. Ray, S.S.; Sinha-Ray, S.; Yarin, A.L.; Pourdeyhimi, B. Theoretical and experimental investigation of physical mechanisms responsible for polymer nanofiber formation in solution blowing. *Polymer (Guildf)* **2015**, *56*, 452–463. [CrossRef]

25. Available online: https://www.oxfordreference.com/view/10.1093/oi/authority.20110803100025554 (accessed on 25 April 2020).

26. Jassim, E.; Abdi, M.A.; Muzychka, Y. Computational fluid dynamics study for flow of natural gas through high-pressure supersonic nozzles: Part 1. Real gas effects and shockwave. *Pet. Sci. Technol.* **2008**, *26*, 1757–1772. [CrossRef]

27. Haddadi, S.A.; Ghaderi, S.; Amini, M.; Ramazani, S.A. ScienceDirect Mechanical and piezoelectric characterizations of electrospun PVDF-nanosilica fibrous scaffolds for biomedical applications. *Mater. Today Proc.* **2018**, *5*, 15710–15716. [CrossRef]

28. Yun, J.S.; Park, C.K.; Jeong, Y.H.; Cho, J.H.; Paik, J.; Yoon, S.H.; Hwang, K. The Fabrication and Characterization of Piezoelectric PZT/PVDF Electrospun Nanofiber Composites. *Nanomater. Nanotechnol.* **2016**, *6*, 62433. [CrossRef]

29. Howling, A.A.; Guittienne, P.; Furno, I. Two-fluid plasma model for radial Langmuir probes as a converging nozzle with sonic choked flow, and sonic passage to supersonic flow. *Phys. Plasmas* **2019**, *26*, 7–10. [CrossRef]

30. Zhang, X.; Wang, N.; Liao, R.; Zhao, H.; Shi, B. Study of mechanical choked Venturi nozzles used for liquid flow controlling. *Flow Meas. Instrum.* **2019**, *65*, 158–165. [CrossRef]

31. Hariharan, P.; Giarra, M.; Reddy, V.; Day, S.W.; Manning, K.B.; Deutsch, S.; Stewart, S.F.C.; Myers, M.R.; Berman, M.R.; Burgreen, G.W.; et al. Multilaboratory particle image velocimetry analysis of the FDA benchmark nozzle model to support validation of computational fluid dynamics simulations. *J. Biomech. Eng.* **2011**, *133*, 1–14. [CrossRef]

32. Okada, A.; Uno, Y.; Onoda, S.; Habib, S. Computational fluid dynamics analysis of working fluid flow and debris movement in wire EDMed kerf. *CIRP Ann.-Manuf. Technol.* **2009**, *58*, 209–212. [CrossRef]

33. Zhang, Z.; Tian, L.; Tong, L.; Chen, Y. Choked flow characteristics of subcritical refrigerant flowing through converging-diverging nozzles. *Entropy* **2014**, *16*, 5810–5821. [CrossRef]

 *polymers*

Article

Modeling the Full Time-Dependent Phenomenology of Filled Rubber for Use in Anti-Vibration Design

Francesca Carleo [1], Jan Plagge [2], Roly Whear [3], James Busfield [1,*] and Manfred Klüppel [2]

[1] School of Engineering and Materials Science, Queen Mary University of London, Mile End Road, London E1 4NS, UK; f.carleo@qmul.ac.uk

[2] Deutsches Institut für Kautschuktechnologie e.V., Eupener Str. 33, 30519 Hannover, Germany; jan.plagge@dikautschuk.de (J.P.); manfred.klueppel@dikautschuk.de (M.K.)

[3] Jaguar Land Rover, Banbury Road, Gaydon CV35 0RR, UK; rwhear1@jaguarlandrover.com

* Correspondence: j.busfield@qmul.ac.uk; Tel.: +44-(0)20-7882-8866

Received: 16 March 2020; Accepted: 1 April 2020; Published: 6 April 2020

Abstract: Component design of rubber-based anti-vibration devices remains a challenge, since there is a lack of predictive models in the typical regimes encountered by anti-vibration devices that are deformed to medium dynamic strains (0.5 to 3.5) at medium strain rates (0.5/s to 10/s). An approach is proposed that demonstrates all non-linear viscoelastic effects such as hysteresis and cyclic stress softening. As it is based on a free-energy, it is fast and easily implementable. The fitting parameters behave meaningfully when changing the filler volume fraction. The model was implemented for use in the commercial finite element software ABAQUS. Examples of how to fit experimental data and simulations for a variety of carbon black filled natural rubber compounds are presented.

Keywords: viscoelastic behaviour; modelling; elastomer; cyclic softening

1. Introduction

Suspension is one of the most important systems affecting a vehicle's ride quality, and it is, therefore, a key factor in determining a vehicle's performance. The modelling of bushing force in the automotive industry plays an important role in predicting the dynamic behaviour of the suspension system. The viscous behaviour of elastomers makes the use of rubber components essential to reduce the level of vibrations that are transmitted to the passengers in the vehicle. In particular, natural rubber is the material of choice for these types of engineering applications. For automotive suspension components, carbon black is also typically used as a filler to improve the rubber's mechanical properties such as the fatigue behaviour or the stiffness. For bushing applications, quasi-static deformations up to large strains are of major interest. In these particular conditions, the rubber exhibits non-linear viscoelastic behaviour such as the Mullins Effect, cyclic stress softening, hysteresis and induced anisotropy [1–5]. This characteristic response is related to the molecular microstructure, but it is not yet totally understood [6]. Furthermore, these effects are also present in unfilled strain-crystallizing elastomers as natural rubber [7–10]. Modelling of filled rubber introduces difficulties related to its nonlinear and incompressible behaviour. Many models try to capture this behavior from a range of different perspectives. The multiscale nature of polymers is reflected by different computational methods for specific length and time scales: quantum ($\sim 10^{-10}$ m, $\sim 10^{-12}$ s), atomistic ($\sim 10^{-9}$ m, $\sim 10^{-9}$–10^{-6} s), mesoscopic ($\sim 10^{-6}$ m, $\sim 10^{-6}$–10^{-3} s) or macroscopic scale ($\sim 10^{-3}$ m, ~ 1 s) [11–13]. At the top level, phenomenological macroscopic models can roughly be classified into three categories [14] damage models, rheological models with serial and parallel combination of elastic and viscous elements and the constitutive equations based on a rubber elasticity model. Damage models [15–17] are unable to differentiate between an unloading and a subsequent reloading with one consequence being that they cannot predict the cyclic stress softening phenomenon. Essentially, there is a limitation in fitting the experimental behaviour as a limitation of the background theory. Rheological Framework models are models with elastic and viscous components. The Parallel Rheological Framework ([18,19]) is capable of reproducing the full range of nonlinear viscoelastic effects that are under examination, but it requires a large number of parameters that are very sensitive to the input test data. The dynamic flocculation model (DFM) assumes the breakdown and reformation of filler aggregates as the main mechanism [20]. Other physically-based works focus on the binding or sliding of polymer chains on the fillers surface [21]. Recent Molecular Dynamics (MD) simulations indicate that both phenomena may play a role [22,23].

Most of these models include softening effects using non-time dependent mathematics, e.g., by softening the material based on maximum strain or stress measures. In contrast, experiments clearly show that softening proceeds by repeatedly stretching the material to a predefined load (cyclic stress softening) or by holding it at constant stretch (stress relaxation) [24]. More specifically, stress decreases logarithmically [25–27]. In a recent review article [14] it was found that no predictive model was able to fully reproduce cyclic stress softening to a satisfying quality. In the next sections a new approach based on a recently proposed model [28] is presented. It is able to account for non-linear elasticity and strain history effects.

2. Theory

2.1. Free Energy Density

The underlying free energy density is derived on the basis of the extended non-affine tube model of rubber elasticity [29,30] which was shown to be the best compromise between fitting quality and number of parameters [31]. In a simplified form it reads

$$W(\tilde{I}_1, \tilde{I}^*) = \frac{G_c}{2} \frac{\tilde{I}_1}{1 - \frac{1}{n}\tilde{I}_1} + 2G_e\,\tilde{I}^* \tag{1}$$

with

$$\tilde{I}_1 = I_1 - 3 = \lambda_1^2 + \lambda_2^2 + \lambda_3^2 - 3 \tag{2}$$

$$\tilde{I}^* = I^* - 3 = \lambda_1^{-1} + \lambda_2^{-1} + \lambda_3^{-1} - 3 \tag{3}$$

being modified invariants of the left Cauchy Green tensor, expressed by the principal stretches λ_i. The straightforward simplifications done to arrive at Equation (1) are outlined in [32]. The parameter G_c is called crosslink modulus and scales the contribution of crosslinks to the mechanical response. Accordingly, G_e scales proportionally to the number of trapped entanglements of the polymer [33]. The third parameter n measures the number of statistical segments between network nodes, which is a measure of elastically effective chain length. For example, natural rubber has a segment length of about 0.934 nm [34], corresponding to roughly two isoprene units. Fillers are introduced by assuming that they amplify strain heterogeneously within the matrix by local amplification factors X [28]. To avoid problems arising from frame references the amplification factors directly act on the invariants. In the case of $\tilde{I}_1$ this is reasonable, as it represents the norm of a hypothetical length within the material. The amplified energy density is then calculated as a superposition of differently amplified domains

$$W_X(\tilde{I}_1, X_{\max}, X_{\min}) = \int_{X_{\min}}^{X_{\max}} \mathrm{d}X\, P(X)\, W(X\tilde{I}_1, X\tilde{I}^*) \tag{4}$$

where the distribution of amplification factors is given by

$$P(X) = (X + C)^{-\chi} \cdot \frac{\chi - 1}{(X_{\min} + C)^{1-\chi} - (X_{\max} + C)^{1-\chi}} \tag{5}$$

The parameter χ gives the width of the distribution and C defines a plateau value around $X = 1$. For $C = 0$ Equation (5) is equivalent to its equivalent in [28]. It is normalized to the interval $[X_{\min}, X_{\max}]$, which is motivated from conservation of the number of rubber-filler structures. Carrying out the integral given by Equation (4) generates hypergeometrical functions. Additive splitting of the integrand, as is explained in [32], gives a good approximation containing only elementary functions:

$$\begin{aligned}
W_X(X_{\max}, X_{\min}) &\approx \tfrac{1}{2} \frac{1}{(X_{\min}+C)^{1-\chi} - (X_{\max}+C)^{1-\chi}} \cdot \\
&\left[\tfrac{G_c \tilde{I}_1 + 4\,G_e \tilde{I}^*}{\chi - 2} \left((C + X_{\min})^{1-\chi}(C + (\chi - 1)\,X_{\min}) - (C + X_{\max})^{1-\chi}(C + (\chi - 1)\,X_{\max}) \right) \right. \\
&\left. + G_c\, n^2(\chi - 1)\,\tilde{I}_1^{\chi - 1} \left(C\,\tilde{I}_1 + n \right)^{-\chi} \log\left(\tfrac{\tilde{I}_1 X_{\min} - n}{\tilde{I}_1 X_{\max} - n} \right) \right]
\end{aligned} \tag{6}$$

From Equation (6) the hyperelastic stresses can be derived, if $X_{\min}$ and $X_{\max}$ are known. The minimum amplification factor is set to $X_{\min} = 1$, which corresponds to the assumption that there are non-amplified domains (e.g., without filler) within the material.

2.2. Stress Softening

In [28] the maximum amplification factor is defined to be a monotonically decreasing function of the all-time maximum of the modified first invariant $\tilde{I}_1$. This naturally introduces immediate stress softening if the material surpasses the previous maximum strain. As described in the introduction stress softening is not immediate, but proceeds logarithmically or according to a slow powerlaw. Generally, logarithmic relaxation in stress $f \sim - \log t$ can be generated by differential equations of the form $df/dt \sim - \exp|f|$. From a physical point of view this can be understood in terms of force-induced hopping over a potential barrier, as formulated by Kramers [35]. It is reasonable to assume that softening is predominantly happening in the most stretched domains. A crude approximation of the stress-scaling in this regime gives

$$f \sim \frac{1}{\langle \lambda \rangle} W(X\tilde{I}_1, X\tilde{I}^*) \approx \frac{1}{\sqrt{\tilde{I}_1}} \frac{\tilde{I}_1 X_{\max}}{1 - \tfrac{1}{n}\tilde{I}_1 X_{\max}} \tag{7}$$

where the entanglement part of Equation (1) was neglected, because its influence is small at high strains. The prefactor $\langle \lambda \rangle = \sqrt{\tilde{I}_1}$ is a frame independent measure of the systems stretch. Heuristically, stress is also directly scaled by the amplification factor, such that $df/dt \sim dX_{\max}/dt$. Using again the logarithm-generating differential equation $df/dt \sim - \exp|f|$ and introducing scaling constants c_1 and c_2 this gives

$$\frac{dX_{\max}}{dt} = - \exp\left(-c_2 + \frac{1}{c_1} \frac{\sqrt{\tilde{I}_1}\, X_{\max}}{1 - \tfrac{1}{n}\tilde{I}_1 X_{\max}} \right) \cdot \frac{1}{s} \tag{8}$$

The timescale of zero-load softening given as $\tau_{\text{soft}} = \exp(c_2)\,s$ was put in the exponential for numerical reasons. In practice, it should be sufficiently long that no significant softening happens on the timescale of the simulation, if no load is imposed. This means that for $\tilde{I}_1 \to 0$ the result of Equation (8) is usually small and the maximum amplification factor $X_{\max}$ stays almost constant. When the model approaches divergence at $\tilde{I}_1 X_{\max}/n \to 1$ the second summand inside the exponential increases and induces a decrease of $X_{\max}$. It shall be noted that a more careful derivation was carried out in [32] which relates c_1 and c_2 to temperature and allows the modeling of Mullins effect recovery [6] at elevated temperatures.

2.3. Viscoelasticity

Hysteresis is modeled by a single Maxwell element. The problem treated in this work requires only a narrow range of strain rates such that this procedure is justified. Viscoelastic stress is evolved according to the differential equation

$$\frac{d\sigma_{vis}}{dt} = -\frac{1}{\tau_c}\sigma_{vis} + \frac{d\sigma_{el}}{dt} \tag{9}$$

where σ_{vis} is the viscoelastic stress contribution, τ_c is the relaxation time scale, and σ_{el} is the elastic Cauchy stress derived from the free energy density given by Equation (6).

2.4. Implementation

The model was implemented in Matlab for the hypothetical condition of fully incompressible material, considering one material point loaded cyclically in the uniaxial direction and in a compressible full three dimensional form for the finite element analysis. For incompressible materials $J = I_3 = 1$ and the Cauchy stress is related to the free energy density:

$$\sigma_{el} = -p\mathbf{I} + 2\left(\frac{\partial W_X}{\partial I_1} + \frac{\partial W_X}{\partial I^*}\right)\mathbf{B} \tag{10}$$

The scalar p is the hydrostatic pressure (an indeterminate Lagrange multiplier) that can be determined only from the boundary conditions. $\mathbf{B}$ is the left Cauchy-Green deformation tensor. For compressible materials it is useful to split the deformation into distortional and volumetric (dilational) parts via a multiplicative decomposition.

$$\mathbf{F} = J^{1/3}\bar{\mathbf{F}} \tag{11}$$

$$\mathbf{B} = J^{2/3}\bar{\mathbf{B}} \tag{12}$$

$$I_1 = J^{2/3}\bar{I}_1 \tag{13}$$

The $J^{1/3}$ and $J^{2/3}$ terms are related to the volume changes, while the $\bar{\bullet}$ terms are a modified gradient and strain related to the distortional deformations. These decompositions express the stored energy as $W(\bar{I}_1(I_1, J), \bar{I}^*(I^*, J), J)$, that can be decomposed into volumetric and isochoric (distortional) responses (Equation (14)) and give the Cauchy stress [36] as in Equation (15)

$$W(\mathbf{B}) = W_{vol}(J) + W_{iso}(\bar{\mathbf{B}}). \tag{14}$$

$$\sigma_{el}(\bar{I}_1, \bar{I}^*, J) = \frac{2}{J}\left[\frac{\partial W_X}{\partial \bar{I}_1} + \frac{\partial W_X}{\partial \bar{I}^*}\right]\bar{\mathbf{B}} + k(J-1). \tag{15}$$

The single Prony series element was integrated explicitly and the total stress is defined as:

$$\sigma_{tot} = (1-\phi)\sigma_{el} + \phi\sigma_{vis}. \tag{16}$$

where ϕ is a fitting constant which is assumed to increase with filler volume fraction. Figure 1 shows the algorithm developed to implement the model as a user-defined material subroutine (VUMAT) coded in Fortran for Abaqus/Explicit.

The model requires 10 parameters, plus the bulk modulus in case of compressible material.

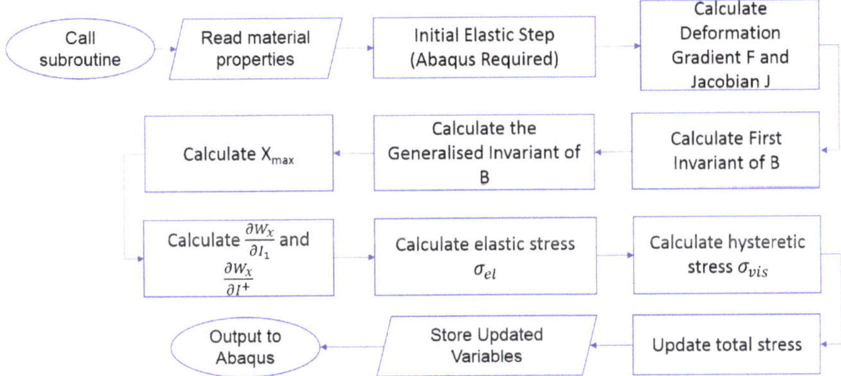

Figure 1. VUMAT Algorithm.

3. Materials and Experiments

Seven compounds supplied by TARRC (Tun Abdul Razak Research Centre, Brickendonbury, Hertford, UK), were examined. Each compound contained a different volume fraction of carbon black (FEF N550) that had been mixed into natural rubber (NR, SMR CV60). The compound formulations in phr (part per hundred of rubber by mass) are given in Table 1.

Table 1. Compound formulation; phr (part per hundred of rubber by mass).

-	NR2	NR10	NR20	NR30	NR40	NR50	NR60
Natural Rubber, SMR CV60	100	100	100	100	100	100	100
Carbon Black, FEF N550	2	10	20	30	40	50	60
Process oil, 410	-	1	2	3	4	5	6
Zinc oxide	5	5	5	5	5	5	5
Stearic acid	2	2	2	2	2	2	2
Antioxidant/antiozonant, HPPD	3	3	3	3	3	3	3
Antiozonant wax	2	2	2	2	2	2	2
Sulfur	1.5	1.5	1.5	1.5	1.5	1.5	1.5
Accelerator, CBS	1.5	1.5	1.5	1.5	1.5	1.5	1.5
t_{90} (min)	15:16	13:50	11:50	10:30	10:00	9:04	7:10

Characterisation of all this materials under cyclic tensile tests is available in [14] and it is briefly recalled in this section for the sake of completeness. Uniaxial tension tests were conducted using an Electropulse Instron Test Machine. Cyclic loading, unloading and reloading tests were performed using four different strain rates of 0.5/s, 1.5/s, 3/s and 6/s. All the specimens had a gauge length of 12 mm and a cross section of 2 mm $\times$ 0.5 mm. The nominal strain, ε^{NOM}, was determined as the ratio of the axial displacement to the original length of a specimen (12 mm). This test was difficult to conduct because the large strain amplitude and rate approached the limit of the test machine. To reach the desired strain rates, the specimens had to be short. In addition, the strain rate was too fast to be reliably measured using the optical strain measuring device. The tensile force was measured using a 1 kN load cell.

A subset of the results are shown in Figures 2 and 3 for two of the materials. The results display strong nonlinearity, large hysteresis, complex Mullins behaviour (Figures 2 and 3b), cyclic stress relaxation (Figure 3a) and permanent set. Full detail of all the experimental results can be found in [14].

In the range of interest the strain rate effect is not a dominant feature, so in the following analysis it was ignored.

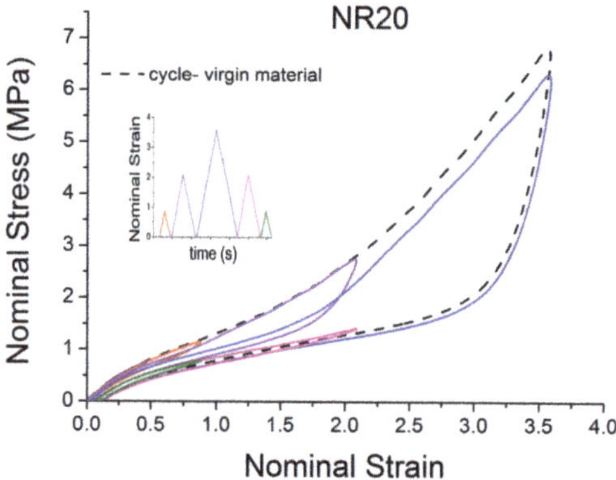

Figure 2. Stress–strain response of natural rubber filled with 20 phr of carbon black (NR20) when subjected to cyclic uniaxial tension. Different behaviour of the material after a different pre-strain is evident.

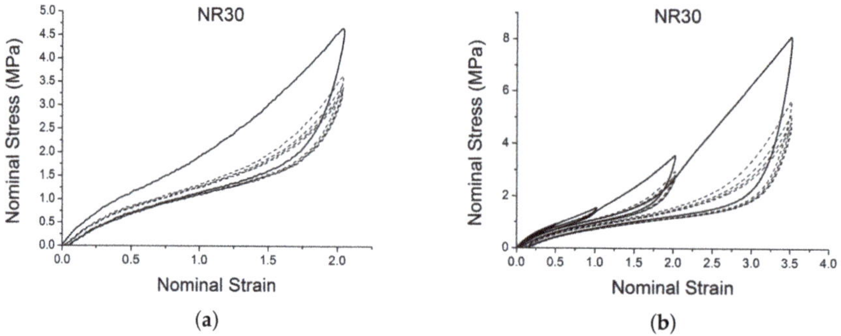

Figure 3. Stress–strain response of natural rubber filled with 30 phr of carbon black (NR30) submitted to cyclic uniaxial tension at 1.5/s. The solid curves are the trends for the first cycle at a given amplitude. The dotted curves are the trends after the initial cycle at given amplitude. (**a**) 5 cycles at the same amplitude, (**b**) 5 cycles at 4 different strain amplitude.

4. Results

4.1. Sensitivity Analysis

This section shows how each parameter effects the output stress. The initial parameters shown in Table 2 are used, it is an arbitrary set. The parameter $X_{max,0} = X_{max}(t = 0)$ is the boundary condition of the differential Equation (8).

Table 2. Material parameters used in the sensitivity analysis and in the VUMAT.

G_c	G_e	ϕ	$X_{max,0}$	χ	n	c_1	c_2	C	τ_c	k
0.557	0.26	0.33	11.27	2.29	21.56	11.20	2.17	−0.78	0.399	100

Figure 4 shows the effect of three of those parameters G_c, G_e, $X_{max,0}$ on the predicted outputs. Figure 5 shows the effect of parameters χ, C and c_1. Figure 6 shows the effect of the parameters c_2, n and τ_c. Figure 7 shows the effect of the parameter ϕ. G_c and G_e, which respectively represent the crosslink modulus and entanglement modulus in the extended tube model, influence the stiffness of the material. The parameter $X_{max,0}$ determines the value of X_{max} before any deformation in the system, and it influences cyclic stress relaxation phenomena. The parameters χ and C characterise the power law distribution of the amplification factor. They influence the response of the material in the loading paths, modifying the the level of stress and the dissipated energy and defining the level of continuum damage due to cyclic stress softening. n, c_1,and c_2 define the relaxation of the maximum amplitude factor to predict the continuum damage, so they operate on the response of the material at a strain higher than 1 and envelop the cyclic stress relaxation process. The constant c_2 scales the zero-load relaxation of the material (which is usually very small), and thus determines, together with c_1 the stress threshold to overcome before softening starts. The parameter ϕ balances elastic stresses (including softening) and viscous effects. The timescale τ_c defines the amount of dissipated energy.

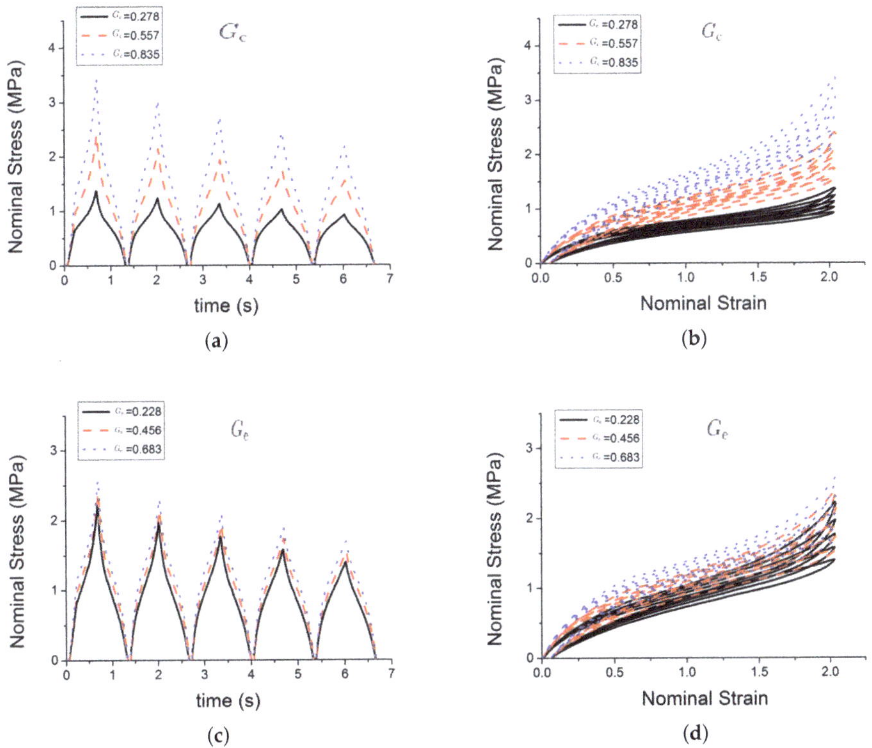

Figure 4. *Cont.*

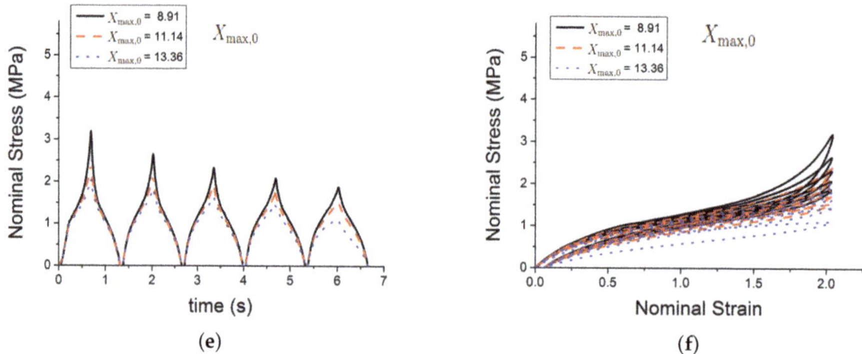

Figure 4. Influence of parameters. (**a**,**b**) show the effect of the parameter G_c; (**c**,**d**) show the effect of the parameter G_e; (**e**,**f**) show the effect of the parameter $X_{\text{max},0}$.

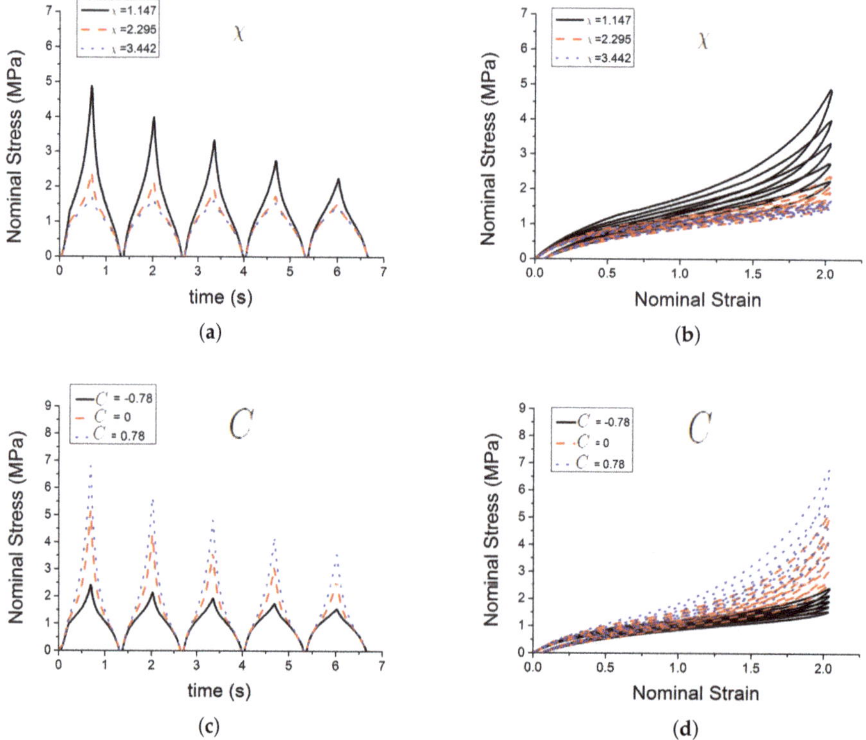

Figure 5. *Cont.*

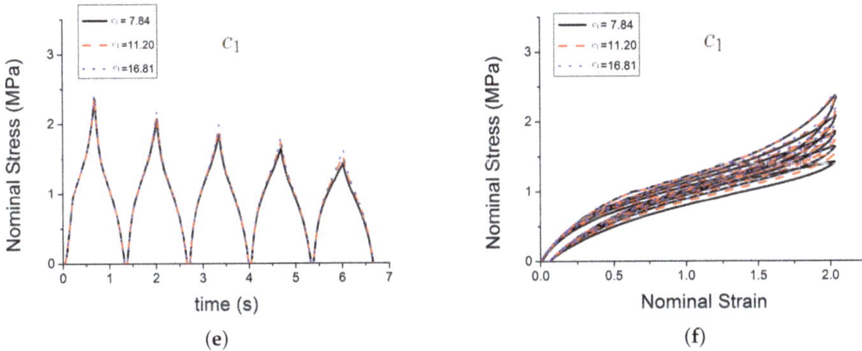

Figure 5. Influence of parameters. (**a**,**b**) show the effect of the parameter χ; (**c**,**d**) show the effect of the parameter C; and, (**e**,**f**) show the effect of the parameter c_1.

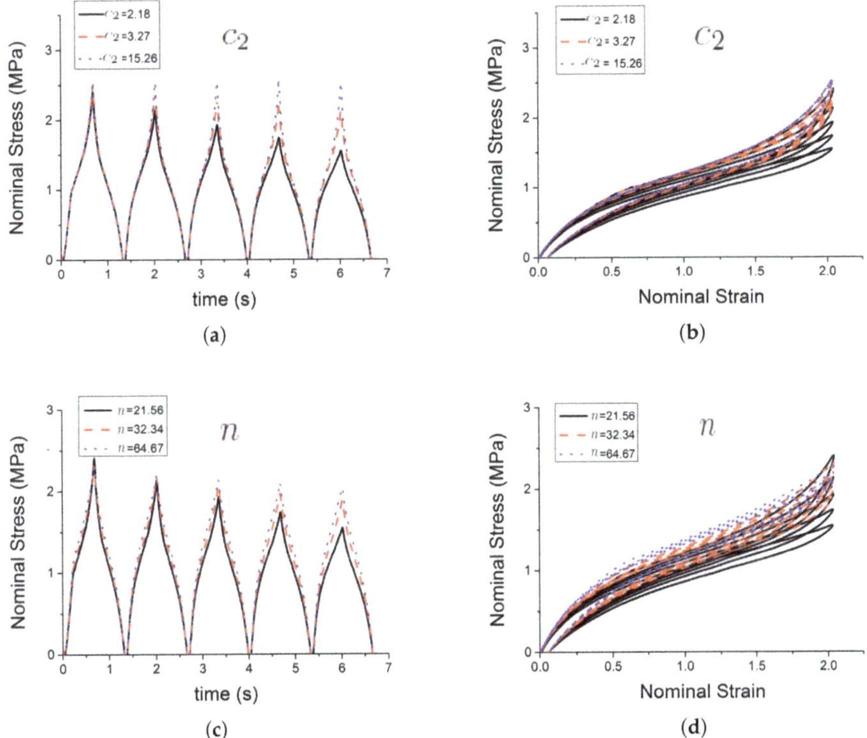

Figure 6. *Cont.*

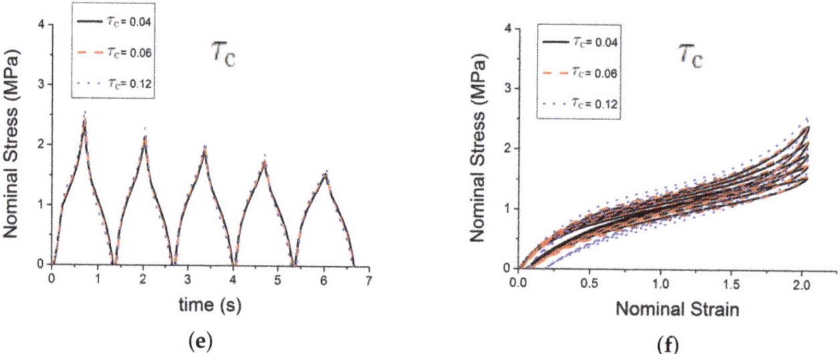

Figure 6. Influence of parameters. (**a**,**b**) show the effect of the parameter c_2; (**c**,**d**) show the effect of the parameter n; and, (**e**,**f**) show the effect of the parameter τ_c.

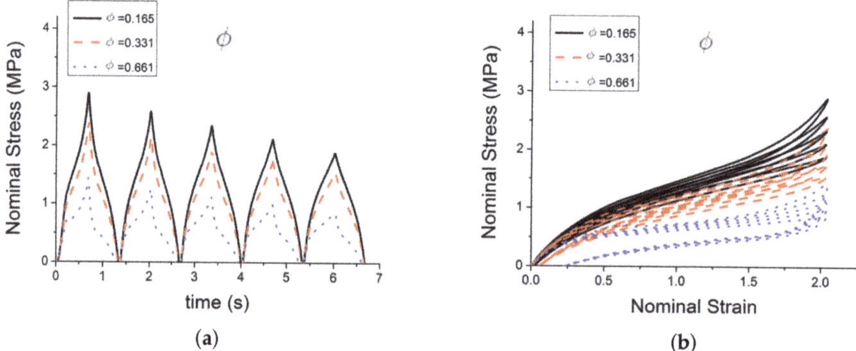

Figure 7. Influence of parameters. (**a**,**b**) show the effect of the parameter ϕ.

4.2. Model Fit to Experimental Data

Model parameters were determined by minimising the mean square error between the model predictions and experimental data. Figures 8a–c show comparisons of model predictions for a selection of materials and loading configurations. For compounds with different amounts of carbon black, the model accurately reproduces the initial loading curve, relaxation cycles, the Mullins effect and the pre-strain effect.

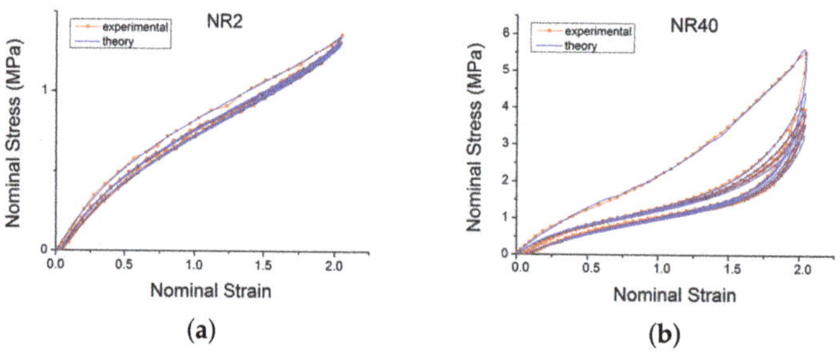

Figure 8. *Cont.*

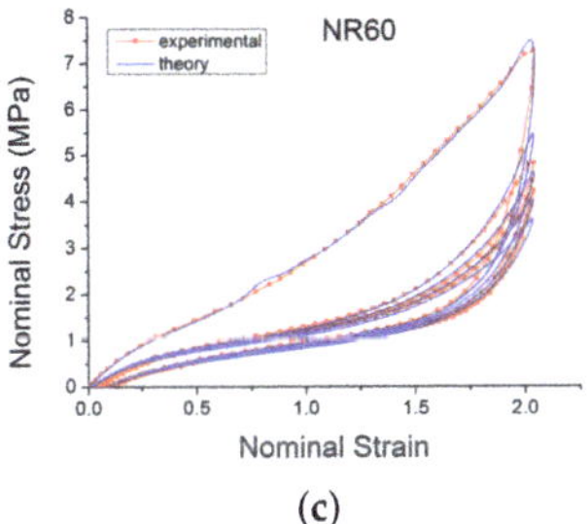

(c)

Figure 8. Comparison of the new model with experimental data loaded with cyclic uniaxial test with uniform maximum strain amplitude [14]. Subfigures (**a**–**c**) show respectively the behaviour for NR2, NR40 and NR60.

4.3. Effect of Carbon Black Content on the Parameters Used in the New Model

The uniaxial tests, where the compound were stretched cyclically with the step up of maximum strain amplitude reached, represent a perfect test to highlight all the non linear viscoelastic behaviour under examination (Mullins effect, cyclic stress softening and permanent set) (Figure 9). The seven compounds were fitted starting from the material with the smaller amount of filler and using the new set of parameters as the guess set for the following compound. The best-fit parameters for different compounds loaded with this strain history show a robust trend with the true carbon black filler volume.

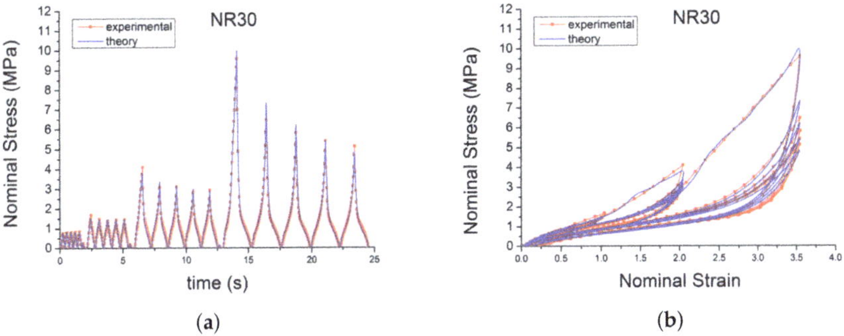

(a) (b)

Figure 9. Comparison of the new model with experimental data for NR30 loaded with cyclic uniaxial stress with different maximum strain amplitude in step up. Subfigure (**a**) shows the Nominal Stress versus time. Subfigure (**b**) shows the Nominal Stress versus Nominal Strain.

The trend of the parameters with the true Carbon Black (CB) filler volume is presented in Figure 10.

- G_c is the modulus resulting from network crosslink and it is relatively constant with the CB volume fraction. It is found to be in a typical range for Natural Rubber vulcanizates. There is a drop in crosslink density at the highest filler loading, whose origin is not clear and may be due to experimental uncertainties and/or parameter correlations.
- G_e, the modulus resulting from network entanglement, is monotonically increasing with the CB concentration. This is surprising and lacks a clear explanation up to now. Probably this parameter somehow captures the increasing low-strain stiffness (Payne effect), even though it was not designed to fulfil this purpose.
- ϕ, represents the scaling of the elastic and inelastic part, is proportional to the true filler volume fraction.

- n, the distance between the network nodes is in a typical range, too. It is constant for the samples containing less than 30 vol. % of carbon black and then slightly decreases, roughly in accordance with the small drop in crosslink modulus at these filler concentrations.
- χ, the exponent of the amplification factor distribution, is approximately constant.
- τ_c, is decreasing with filler volume fraction.
- $X_{max,0}$ is approximately constant with filler volume fraction.
- C is increasing. The parameter was introduced primarily to modify the shape of the virgin loading curve. A value $k > 0$ generates a rather linear curve, as is observed for many highly filled compounds.
- c_2 decreases with volume fraction and appears to asymptotic. From Equation (8) it can be seen that exp c_2 defines the timescale on which $X_{max,0}$ relaxes without load. The value obtained from fitting here create an optimal model for the timescale of the fit, but it may fail for longer simulation times. An optimal determination of c_2 and c_1 requires a stress relaxation characterisation in the fitting data.
- c_1 increases modestly with volume fraction, probably representing that the rubber-filler structures to be broken down during softening become more rigid at higher filler loadings.

Moreover, it can be seen that the parameters G_e, C, c_1, ϕ and τ_c show some kind of discontinuity in the range of 9–13 vol. % carbon black loading. This roughly corresponds to the percolation threshold of the filler network inside the polymer matrix [37].

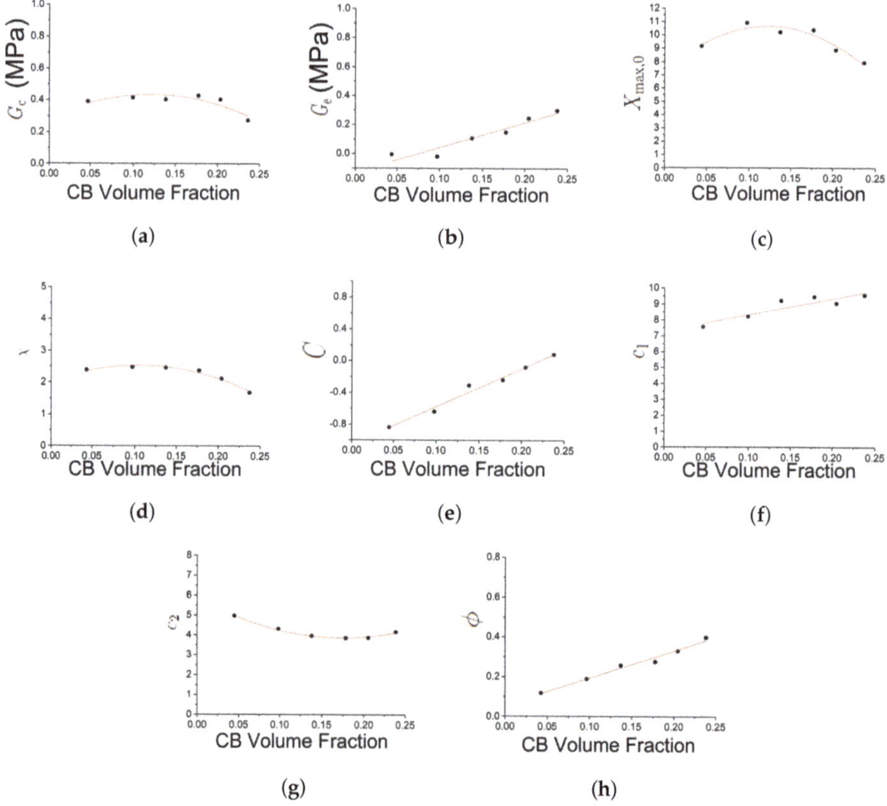

Figure 10. *Cont.*

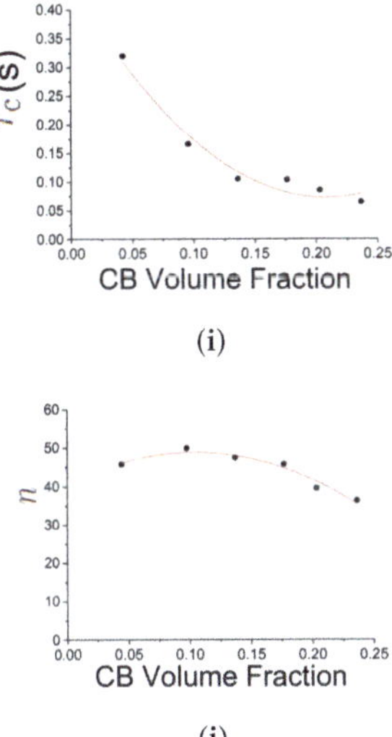

(i)

(j)

Figure 10. Trend of he parameters with the true carbon black filler volume V_f. Subfigures (**a–j**) show respectively the trend of G_c, G_e, $X_{max,0}$, χ, C, c_1, c_2, ϕ, τ_c and n.

4.4. Finite Element Analysis

Benchmark Tests

This section shows the benchmark tests on a single element model with a simple prescribed traction loading. The model is a cube of size 0.5 mm and is meshed with 1 linear brick element with reduced integration (C3D8R). A velocity of 1 mm/s is applied in direction 1 (along the x-axis). The element was loaded for 5 cycles at constant max strain amplitude using 10 steps. The velocity was positive ($v = 1$ mm/s) in the odd steps to stretch the element and negative ($v = -1$ mm/s) in the even steps to unload the element. The job was run using double precision. Material parameters are shown in Table 2. Figures 11 and 12 show the typical output for two different cyclic loading path.

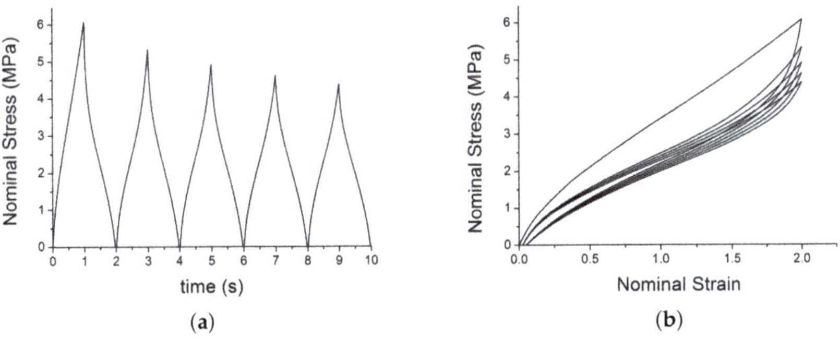

(a)

(b)

Figure 11. Test 1: VUMAT output: (**a**) shows the nominal stress versus time; (**b**) shows the nominal strain–stress path with the set of parameters in Table 2.

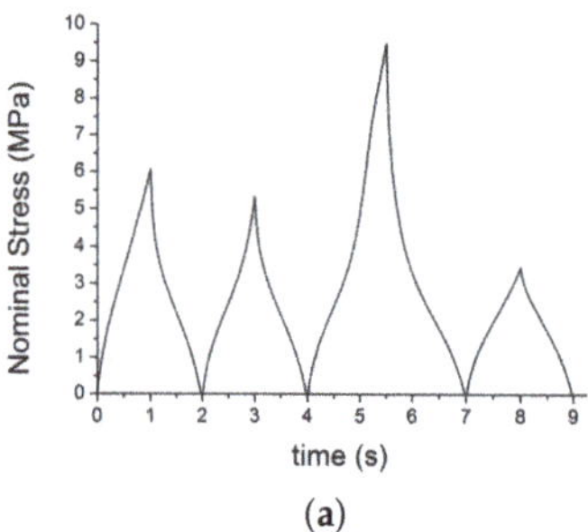
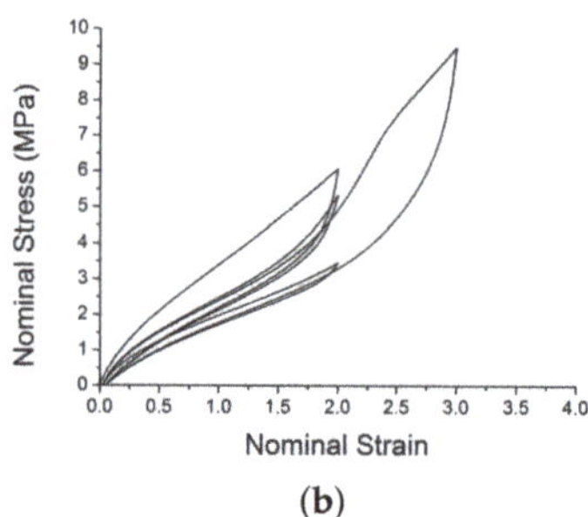

(a)   (b)

Figure 12. Test 2: VUMAT output: (**a**) shows the nominal stress; (**b**) shows the nominal strain–stress path with the set of parameters in Table 2.

5. Discussion and Conclusions

The new model is based on physical rubber elasticity considerations. The total stress is the weighted sum of two contributions: an amplified, simplified extended tube model, and a hysteretic part. The hysteretic part was implemented using a single Maxwell/Prony element. For compounds with different CB volume fraction the model reproduces the non-linear behaviour under examination: Mullins effect, cyclic stress relaxation, permanent set, and hysteresis. The model requires 10 material parameters. The best-fit sets of parameters show a linear or a polynomial trend with the CB volume fraction. The sensitivity study shows the effect of altering each of the parameters. The FE implementation of the new model reproduces the expected behaviour. Finite Element Analysis was conducted on single-element models and on a realistic component. This preliminary investigation demonstrates that this model is stable when incorporated into a finite element model. The VUMAT works properly with different structures, and it is able to predict the desired non–linear viscoelastic behaviour. The new model was already implemented in the Jaguar Land Rover (JLR) suspension tool. The first job carried out a single cycle at 20 kN on a suspension loaded radially. Preliminary results are reported in Figure 13. Further studies could focus on the modeling of a broader range of deformation rates potentially requiring the use of more Prony elements. Moreover, the idea of a static hysteresis, as implemented using an intrinsic time in [28], deserves further investigation.

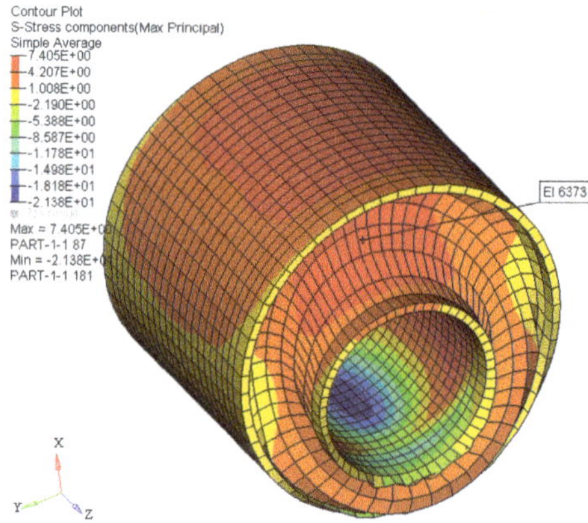

Figure 13. JLR toolbox: preliminary results.

Author Contributions: F.C. developed the curve fitting algorithm, coding and implementation and performed simulation. J.P. developed the non linear elastic model. F.C. and J.P. prepared the first draft of the manuscript. R.W. sponsored the project. J.B. and M.K. supervised the project throughout and contributed to detailed editing of the manuscript. All authors have read and approved the final version of the manuscript.

Funding: The project was funded by Jaguar Land Rover, Banbury Road, Gaydon CV35 0RR, UK.

Acknowledgments: The author would like to acknowledge the financial support provided by Jaguar Land Rover.

Conflicts of Interest: The authors declare no conflict of interest.

Abbreviations

The following abbreviations are used in this manuscript:

DFM   Dynamic Flocculation Model
MD   Molecular Dynamics
TARRC   Tun Abdul Razac Reseach Centre
NR   Natural Rubber
SMR   Standard Malaysian Rubber
FEF   Fast Extruding Furnace
PHR   Part per Hundred Rubber
CB   Carbon Black
JLR   Jaguar Land Rover

References

1. Li, Z.; Wang, Y.; Li, X.; Li, Z.; Wang, Y. Experimental investigation and constitutive modeling of uncured carbon black filled rubber at different strain rates. *Polym. Test.* **2019**, *75*, 117–126. [CrossRef]

2. Machado, G.; Chagnon, G.; Favier, D. Induced anisotropy by the Mullins effect in filled silicone rubber. *Mech. Mater.* **2012**, *50*, 70–80. [CrossRef]

3. Luo, R.K.; Peng, L.M.; Wu, X.; Mortel, W.J. Mullins effect modelling and experiment for anti-vibration systems. *Polym. Test.* **2014**, *40*, 304–312. [CrossRef]

4. Wang, S.; Chester, S.A. Modeling thermal recovery of the Mullins effect. *Mech. Mater.* **2018**, *126*, 88–98. [CrossRef]

5. Marckmann, G.; Chagnon, G.; Le Saux, M.; Charrier, P. Experimental investigation and theoretical modelling of induced anisotropy during stress-softening of rubber. *Int. J. Solids Struct.* **2016**, *97*, 554–565. [CrossRef]

6. Diani, J.; Fayolle, B.; Gilormini, P. A review on the Mullins effect. *Eur. Polym. J.* **2009**, *45*, 601–612. [CrossRef]

7. Gros, A.; Verron, E.; Huneau, B. A physically-based model for strain-induced crystallization in natural rubber. Part II: Derivation of the mechanical model. *J. Mech. Phys. Solids* **2019**, *125*, 255–275. [CrossRef]

8. Behnke, R.; Berger, T.; Kaliske, M. Numerical modeling of time-and temperature-dependent strain-induced crystallization in rubber. *Int. J. Solids Struct.* **2018**, *141*, 15–34. [CrossRef]

9. Plagge, J.; Klüppel, M. Determining strain-induced crystallization of natural rubber composites by combined thermography and stress-strain measurements. *Polym. Test.* **2018**, *66*, 87–93, doi:10.1016/j.polymertesting.2017.12.021. [CrossRef]

10. Huneau, B. Strain-induced crystallization of natural rubber: A review of X-ray diffraction investigations. *Rubber Chem. Technol.* **2011**, *84*, 425–452. [CrossRef]

11. Horstemeyer, M.F. Multiscale modeling: A review. In *Practical Aspects of Computational Chemistry*; Springer: Berlin, Germany, 2009; pp. 87–135.

12. Gooneie, A.; Schuschnigg, S.; Holzer, C. A review of multiscale computational methods in polymeric materials. *Polymers* **2017**, *9*, 16. [CrossRef] [PubMed]

13. Shojaei, A.; Li, G. Viscoplasticity analysis of semicrystalline polymers: A multiscale approach within micromechanics framework. *Int. J. Plast.* **2013**, *42*, 31–49. [CrossRef]

14. Carleo, F.; Barbieri, E.; Whear, R.; Busfield, J. Limitations of viscoelastic constitutive models for carbon-black reinforced rubber in medium dynamic strains and medium strain rates. *Polymers* **2018**, *10*, 988. [CrossRef] [PubMed]

15. Ogden, R.; Roxburgh, D. A pseudo–elastic model for the Mullins effect in filled rubber. In *Proceedings of the Royal Society of London A: Mathematical, Physical and Engineering Sciences*; The Royal Society: London, UK, 1999; Volume 455, pp. 2861–2877.

16. Dorfmann, A.; Ogden, R. A pseudo-elastic model for loading, partial unloading and reloading of particle-reinforced rubber. *Int. J. Solids Struct.* **2003**, *40*, 2699–2714. [CrossRef]

17. Wrubleski, E.G.M.; Marczak, R.J. A modification of hyperelastic incompressible constitutive models to include non-conservative effects. In Proceedings of the 22nd International Congress of Mechanical Engineering (COBEM 2013), Ribeirão Preto, Brazil, 3–7 November 2013; pp. 2176–5480.

18. Hurtado, J.; Lapczyk, I.; Govindarajan, S. Parallel rheological framework to model non-linear viscoelasticity, permanent set, and Mullins effect in elastomers. *Const. Model. Rubber VIII* **2013**, *95*. [CrossRef]

19. Nandi, B.; Dalrymple, T.; Yao, J.; Lapczyk, I. Importance of capturing non-linear viscoelastic material behavior in tire rolling simulations. In Proceedings of the 33rd Annual Meeting and Conference on the Tire Science and Technology, At Akron, OH, USA, 8–10 September 2014.

20. Lorenz, H.; Klüppel, M.; Heinrich, G. Microstructure-based modelling and FE implementation of filler-induced stress softening and hysteresis of reinforced rubbers. *ZAMM* **2012**, *92*, 608–631. [CrossRef]

21. Khiêm, V.N.; Itskov, M. An averaging based tube model for deformation induced anisotropic stress softening of filled elastomers. *Int. J. Plast.* **2017**, *90*, 96–115. [CrossRef]

22. Kojima, T.; Koishi, M. Influence of Filler Dispersion on Mechanical Behavior with Large-Scale Coarse-Grained Molecular Dynamics Simulation. *Technische Mechanik* **2018**, *38*, 41–54.

23. Hagita, K.; Morita, H.; Doi, M.; Takano, H. Coarse-grained molecular dynamics simulation of filled polymer nanocomposites under uniaxial elongation. *Macromolecules* **2016**, *49*, 1972–1983. [CrossRef]

24. Muhr, A.H. Modeling the Stress-Strain Behavior of Rubber. *Rubber Chem. Technol.* **2005**, *78*, 391–425. [CrossRef]

25. Plagge, J.; Klüppel, M. A hyperelastic physically based model for filled elastomers including continuous damage effects and viscoelasticity. In *Constitutive Models for Rubber X*; CRC Press: Boca Raton, FL, USA, 2017; pp. 559–565.

26. Aoyagi, Y.; Giese, U.; Jungk, J.; Beck, K. The analysis of Aging Processes of EPDM Elastomers using low field NMR with Inverse Laplace Transform and Stress Relaxation Measurements. *KGK-Kautschuk Gummi Kunststoffe* **2018**, *71*, 26–35.

27. Ehrburger-Dolle, F.; Morfin, I.; Bley, F.; Livet, F.; Heinrich, G.; Richter, S.; Piché, L.; Sutton, M. XPCS investigation of the dynamics of filler particles in stretched filled elastomers. *Macromolecules* **2012**, *45*, 8691–8701. [CrossRef]

28. Plagge, J.; Klüppel, M. A physically based model of stress softening and hysteresis of filled rubber including rate-and temperature dependency. *Int. J. Plast.* **2017**, *89*, 173–196. [CrossRef]

29. Kaliske, M.; Heinrich, G. An extended tube-model for rubber elasticity: Statistical-mechanical theory and finite element implementation. *Rubber Chem. Technol.* **1999**, *72*, 602–632. [CrossRef]

30. Vilgis, T.A.; Heinrich, G.; Klüppel, M. *Reinforcement of Polymer Nano-Composites: Theory, Experiments and Applications*; Cambridge University Press: Cambridge, UK, 2009.

31. Marckmann, G.; Verron, E. Comparison of Hyperelastic Models for Rubber-Like Materials. *Rubber Chem. Technol.* **2006**, *79*, 835–858, doi:10.5254/1.3547969. [CrossRef]

32. Plagge, J. On the Reinforcement of Rubber by Fillers and Strain-induced Crystallization. Ph.D. Thesis, Institutionelles Repositorium der Leibniz Universität Hannover, Hannover, Germany, 2018, doi:10.15488/3953. [CrossRef]

33. Klüppel, M.; Heinrich, G. Network structure and mechanical properties of sulfur-cured rubbers. *Macromolecules* **1994**, *27*, 3596–3603. [CrossRef]

34. Mark, J.E. *Physical Properties of Polymers Handbook*; Springer: Berlin, Germany, 2007.

35. Hänggi, P.; Talkner, P.; Borkovec, M. Reaction-rate theory: Fifty years after Kramers. *Rev. Mod. Phys.* **1990**, *62*, 251. [CrossRef]

36. Bergstrom, J.S. *Mechanics of Solid Polymers: Theory and computational modeling*; William Andrew: Burlington, MA, USA, 2015.

37. Heinrich, G.; Klüppel, M. Recent advances in the theory of filler networking in elastomers. In *Filled Elastomers Drug Delivery Systems*; Springer: Berlin, Germany, 2002; pp. 1–44.

Article

Fracture Resistance Analysis of 3D-Printed Polymers

Ali Zolfagharian [1,*], Mohammad Reza Khosravani [2] and Akif Kaynak [1]

[1] School of Engineering, Deakin University, Geelong, Victoria 3216, Australia; akaynak@deakin.edu.au
[2] Chair of Solid Mechanics, University of Siegen, Paul-Bonatz-Str. 9-11, 57068 Siegen, Germany; mohammadreza.khosravani@uni-siegen.de
* Correspondence: a.zolfagharian@deakin.edu.au

Received: 4 January 2020; Accepted: 23 January 2020; Published: 2 February 2020

Abstract: Three-dimensional (3D)-printed parts are an essential subcategory of additive manufacturing with the recent proliferation of research in this area. However, 3D-printed parts fabricated by different techniques differ in terms of microstructure and material properties. Catastrophic failures often occur due to unstable crack propagations and therefore a study of fracture behavior of 3D-printed components is a vital component of engineering design. In this paper, experimental tests and numerical studies of fracture modes are presented. A series of experiments were performed on 3D-printed nylon samples made by fused deposition modeling (FDM) and multi-jet fusion (MJF) to determine the load-carrying capacity of U-notched plates fabricated by two different 3D printing techniques. The equivalent material concept (EMC) was used in conjunction with the J-integral failure criterion to investigate the failure of the notched samples. Numerical simulations indicated that when EMC was combined with the J-integral criterion the experimental results could be predicted successfully for the 3D-printed polymer samples.

Keywords: mechanical fracture; 3D printing; mixed mode fracture; load-carrying capacity

1. Introduction

Three-dimensional printing (3D) has increasingly become one of the prevalent manufacturing methods for the development of tools and functional parts, particularly in the sectors entailing extensive customization, such as medical and aerospace [1–3]. However, there are currently various types of 3D printing techniques, including fused deposition modeling (FDM), stereolithography (SLA), and multi-jet fusion (MJF) that bring quite a number of uncertainties to the prediction of the outcome based on the mechanical engineering concepts [4,5].

Generally, 3D-printed parts fabricated by different techniques differ in terms of strength, stiffness, microstructure and material properties, so that it is necessary to investigate their behaviors through both experimental tests and finite element analysis (FEA) [6–10]. Fused deposition modeling 3D printing has shown promising capabilities to improve the specific mechanical properties of parts—such as fracture toughness—by optimization of extrusion deposition trajectories [11]. It is reported that the filament direction optimization contributed to a large deformation zone leading to "ductile-like behavior", which subsequently caused a slow crack propagation rate [9]. Also, using 3D printing for developing co-continuous composite or interpenetrating phase composite parts proved a significant increase of damage tolerance and fracture toughness compared to conventionally manufacture composite parts [12]. Rationally designed interpenetrating architectures enabled implementing diverse techniques of crack-deflection and crack-bridging to enhance fracture resistance in 3D-printed composites. Mechanical reliability is an essential aspect of various engineering applications. Catastrophic failures often occur due to unstable crack propagations and therefore, study of fracture of components is a vital component of engineering design. In this respect, to address the fracture behavior of 3D-printed plastic components we used FDM and MJF techniques to manufacture notched samples and supported the experimental observations with a failure model.

Pre-existing cracks in different materials can be induced by various circumstances. For instance, in composite materials defects can inadvertently be produced in the service life and also during the manufacturing process [13]. In ductile materials, cracks that originate from surface imperfections or notches often grow slowly, accompanied by significant plastic deformation. However, the rate of crack growth accelerates with the progressively increasing applied stresses [14].

A reliable method for evaluating the notched ductile parts before using them in real applications, i.e., grooves, threads, and shaft holes, was developed through fracture toughness experiments [15]. In earlier studies, mode I and mixed mode conditions have been practically used to determine the fracture behavior of materials [16–18]. Mode I fields are tensile stress normal to the plane of the crack symmetric with respect to the crack line, while the mode II fields are shear stress acting parallel to the plane of the crack and perpendicular to the crack front. Thus, the line ahead of the crack tip has the mixed mode I/II stress intensity factors. In these studies, the finite fracture theory was employed to investigate the fracture of rounded-tip notched samples with various notch radii and inclination angles under tensile loading. However, analysis of the fracture in ductile materials are more critical due to the plastic deformations as well as the stored elastic strain energy around the notch prior to the crack initiation [19]. For the analysis of fracture in notched ductile specimens, elastic-plastic fracture mechanics is usually employed, which is complex and time consuming. Therefore, an alternative method, known as equivalent material concept (EMC), was proposed where the elastic-plastic fracture mechanics analyses are circumvented with an equivalent virtual brittle material to predict ductile failure [20,21].

In this paper, we present an analysis of the fracture of U-notched 3D-printed thermoplastic components made by two different 3D printing techniques. The equivalent material concept was employed to provide a new model in failure prediction. As the J-integral failure criterion is one of the most common brittle failure models used in the study of notched specimens, we investigated whether EMC could be combined with the J-integral failure principle to predict the fracture of U-notched 3D-printed specimens subjected to tensile loading. We present a study of the failure of U-notched 3D-printed rectangular nylon samples made by FDM and MJF under mode I and mixed mode I/II loading regimes within the framework of combined EMC and J-integral criterions. Nylon 12 filament and PA12 nylon powder were used for printing with centrally positioned notches. A series of tensile tests under static loading regime were performed, the results of which compared with that obtained by the EMC model combined with the J-integral criterion.

2. 3D Printing Samples

Two different types of 3D printing technologies were used to 3D print nylon samples. All the samples were first designed in a CAD platform and then saved in ".stl" format. The dogbone-type nylon samples were 3D-printed according to Type I ASTM D638 with a width of 13 mm, thickness of 5 mm and a gauge length of 50 mm as solid 100% infill rectangular plates, weakened by a centrally located bean-shaped slit with two U-shaped ends as shown in Figure 1, were 3D-printed by FDM and MJF with dimensions of $160 \times 40 \times 5$ mm.

In FDM, nylon 12 samples were printed by a Fortus 450mc with nozzle and bed temperatures of 250 °C and 60 °C, respectively, nozzle speed 40 mm/s, infill percentage 100%, layer thickness 0.254, number of contours 1, and raster angle 0°, whereas an HP 3D printer was used to produce the MJF PA12 samples with powder melting 187 °C and infill percentage 100%. The infill percentage was 100% for both samples. The average density of samples were measured using an electronic Qualitest Densimeter SD-200L as 1.019 gr/cm^3 and 0.984 gr/cm^3 for MJF and FDM 3D-printed specimens, respectively.

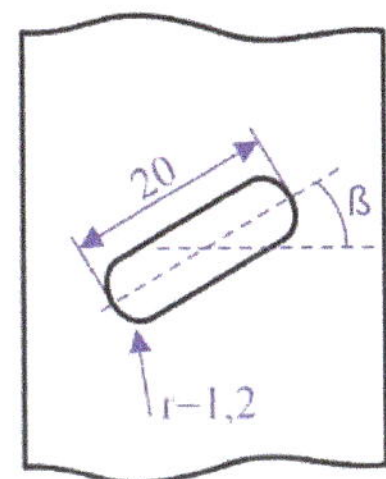

Figure 1. A schematic of centrally located bean-shaped notch with two U-shaped ends (dimensions in mm).

3. Experiments

3.1. Tensile Tests on 3D-Printed Dogbone Samples

In order to determine basic mechanical properties, tensile tests were conducted on 3D-printed dog-bone specimens, as shown in Figure 2. The tests were performed in an Instron 300LX (Instron, High Wycombe, UK) with a crosshead speed of 5 mm/min^{-1}. Tensile strength and modulus of the 3D-printed samples were determined from the stress vs strain graphs shown in Figure 3 and inserted in Table 1. For each 3D printing technique, six samples were tested to obtain the average of the mechanical properties.

The results of tensile tests showed that the average value of the modulus of elasticity of MJF 3D-printed nylon was 780 MPa, whereas the FDM sample had a lower value of 493 MPa. The average value of the percent breaking strain of the FDM and MJF samples were 16 and 13 and the average tensile strength of FDM and MJF samples were 44.8 and 34.9 MPa, respectively. Although FDM samples had lower elastic modulus, they exhibited higher tensile strength and percentage elongation compared to MJF, as well as higher modulus of toughness.

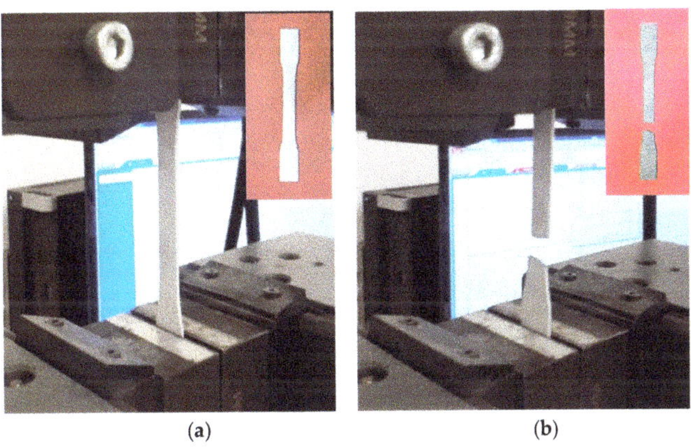

(a) (b)

Figure 2. A dog-bone shaped MJF nylon 3D-printed nylon specimen; before (**a**), and after (**b**) tensile test.

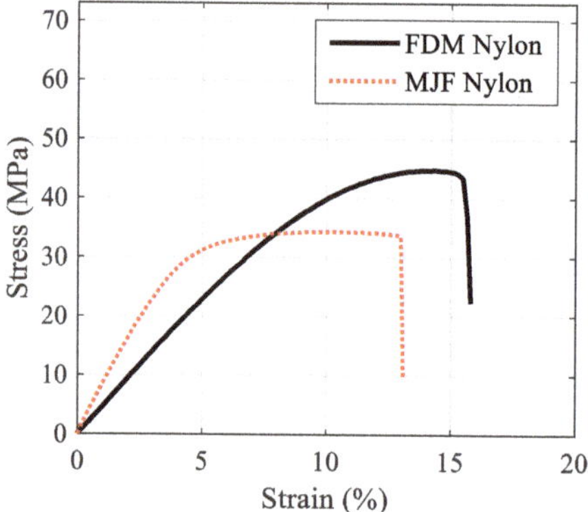

Figure 3. A typical tensile stress vs strain curve for the 3D-printed nylon specimens.

Table 1. Mechanical properties of the 3D-printed nylon specimens.

Material Property	FDM Nylon	MJF Nylon
Elastic modulus, E (GPa)	0.493	0.780
Poisson's ratio, ν	0.39	0.39
Ultimate tensile strength (MPa)	44.79	34.91
Yield stress (MPa), σ_Y	3.61	7.21
Fracture toughness, (MPa m$^{0.5}$)	0.92	0.58
Breaking strain (%), ε_B	16	13
Strain-hardening coefficient, k (MPa)	294.7138	140.9280
Strain-hardening exponent, n (MPa)	0.8437	0.5473
Fracture stress, σ_b (MPa)	43.45	34.03

3.2. Fracture Tests on 3D-Printed Components

To investigate the behavior of samples subjected to various in-plane modes of tensile loading (i.e., mode I and mixed mode I/II loadings) as illustrated in Figure 4a,b. In mode 1 (crack opening) the axial normal stress is applied perpendicular to the plane of the notch; in mode 2 (shear mode), shear stress is applied normal to the crack front, parallel to the crack plane. In mode 3 (tearing mode), the shear stress is applied parallel to the crack front. The samples were designed and fabricated with various inclination angles (β), where β is the angle between the notch bisector line and the horizontal axis. When β is equal to zero, pure mode I loading occurs at the U-shaped ends. As the value of β increases, the in-plane shear strain develops around the U-notch which means with increasing β from zero, the loading mode on the U-notch changes from mode I towards mixed mode I/II. The oblique orientation of the notch with the axial loading gives rise to combined tensile-shear loading conditions. In the current study, we consider specific orientations of the notch at $\beta = 0°$, $\beta = 30°$ and $\beta = 60°$ as well as different notch radii at $r = 1$ mm and $r = 2$ mm so that mode I, mixed mode I/II and notch radius effects could be investigated. The cracks are generated inside the 3D-printed samples where the load is distributed into horizontal and angled walls, to compare required fracture energy values.

The 3D-printed samples prepared by aforementioned approach were axially stressed until the final failure by crack growth using an Instron 300LX (Instron, High Wycombe, UK) machine. Figure 4 shows MJF 3D-printed nylon specimen under tensile test conditions before (Figure 4a) and after fracture (Figure 4b). The tests were conducted under displacement controlled conditions at a constant cross-head speed of 20 mm/min. The photographs of the MJF- and FDM-printed nylon specimens before and after fracture at three different notch orientations and two different radii are shown in Figure 5. In each radius and inclination angle, three samples were tested to obtain the stress distribution and average of the critical load during crack growth.

The representative stress–strain curves of the FDM and MJF 3D-printed nylon specimens at three different notch angles and two different notch radii are demonstrated in Figure 6. In all the cases, the ultimate tensile stress on the samples increased with an increase of inclination angle. The FDM 3D-printed nylon is clearly more ductile as it has shown approximately three times higher strain at break than MJF 3D-printed samples. The pronounced difference in tensile behavior can be attributed to the printing method employed. In the MJF method, a layer of powder is evenly distributed, then a fusing agent is applied followed by application of uniform heat to fuse the powder into an isotropic solid layer, whereas in the case of the FDM the thermoplastic filament is melted as it is positioned to form the layer. The continuous nature of the filament aligned in the direction of the applied stress resulted in better structural integrity than that achieved by fusing the polymer powder, thus resulting in significantly higher elongation at break and toughness. An examination of the close-up photographs of fractured FDM and MJF samples in Figure 7 demonstrates a ductile fracture in the case of FDM, with significant plastic deformation below the fracture surface accompanied by stress whitening. The narrowing of the cross section in the vicinity of the fracture surface indicates extensive stretching (Figure 7a,c). In contrast, close-up images of the MJF samples (Figure 7b,d) do not show any significant plastic deformation prior to failure. Irrespective of the notch orientation (Figures 5 and 7), all MJF samples exhibited brittle behavior with flat fracture surfaces. Considering the tensile data in Table 2, it is evident that the failure load of MJF 3D-printed nylon at both radii is greater than FDM samples for $\beta = 0°$ and 30°. However, this is not the case when the inclination angle increased to 60°. In addition to the effect of crack angles; it was observed that increase in crack radius was associated with reduced critical load in both types of 3D-printed samples.

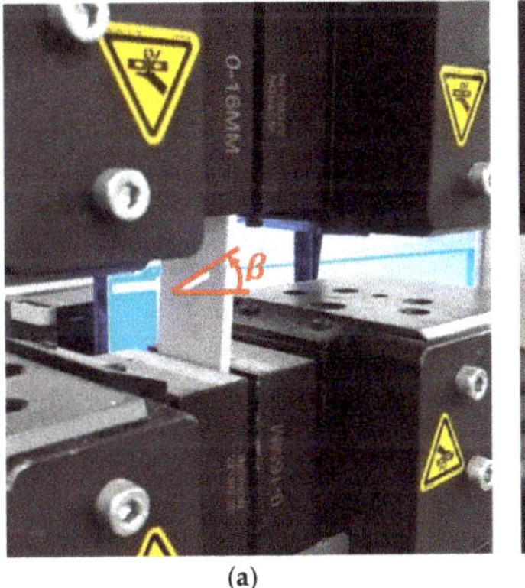

(a)

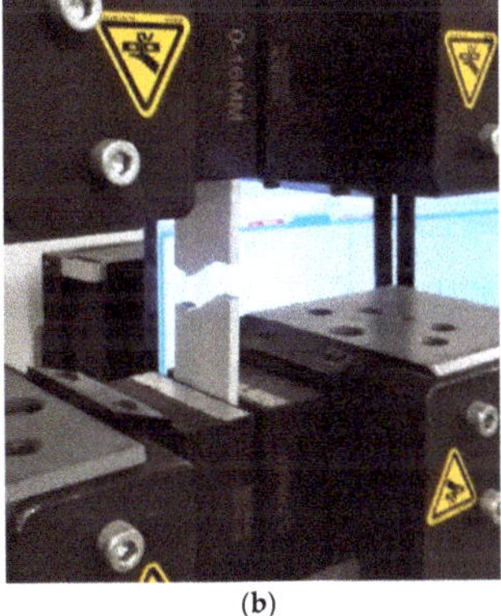

(b)

Figure 4. An MJF nylon 3D-printed specimen under tensile test conditions; (**a**) before (**b**) after fracture.

Table 2. The experimental failure loads of the examined 3D-printed specimens.

Material	β (deg.)	ρ (mm)	P_{av} (N)
FDM Nylon	0	1	5011.8 ± 0.4
		2	4015.1 ± 3.1
	30	1	5667.3 ± 9.5
		2	4968.6 ± 8.9
	60	1	7047.5 ± 4.7
		2	6714.7 ± 3.6
MJF Nylon	0	1	5441.2 ± 5.7
		2	4749.6 ± 2.8
	30	1	5802.8 ± 2.1
		2	5547.7 ± 1.9
	60	1	6520.2 ± 4.1
		2	6219.2 ± 3.8

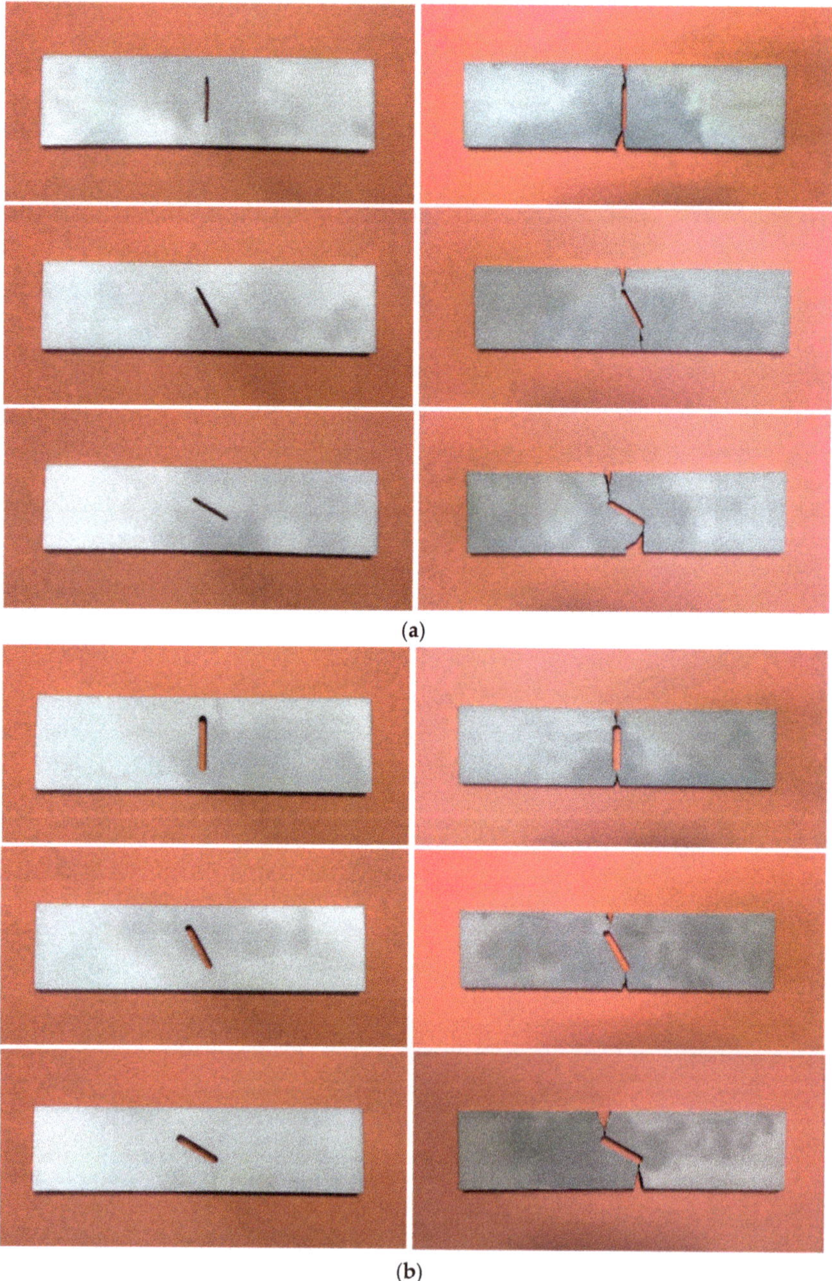

(a)

(b)

Figure 5. *Cont.*

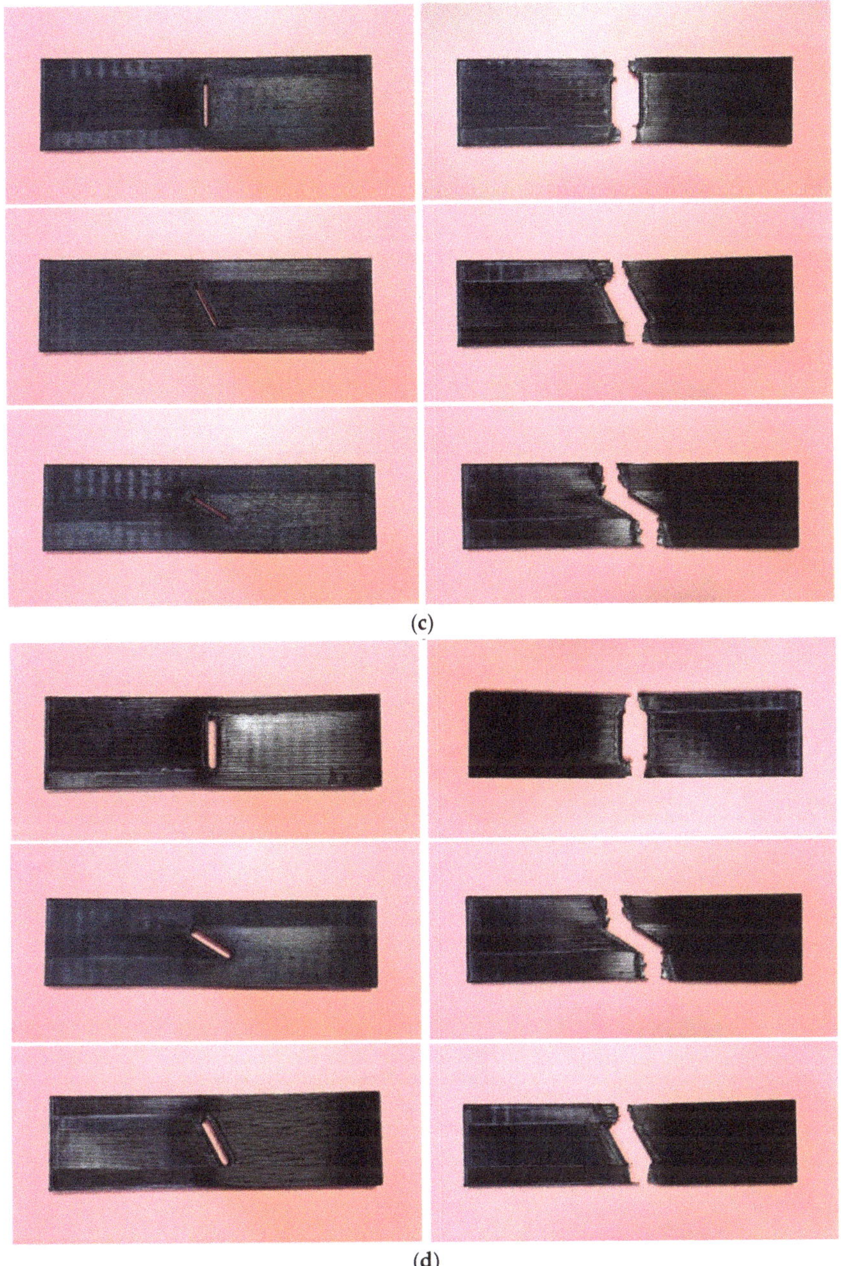

Figure 5. 3D-printed specimens before (left) and after (right) rupture; (**a**) MJF nylon, $r = 1$ mm, (**b**) MJF nylon, $r = 2$ mm, (**c**) FDM nylon, $r = 1$ mm and (**d**) FDM nylon, $r = 2$ mm.

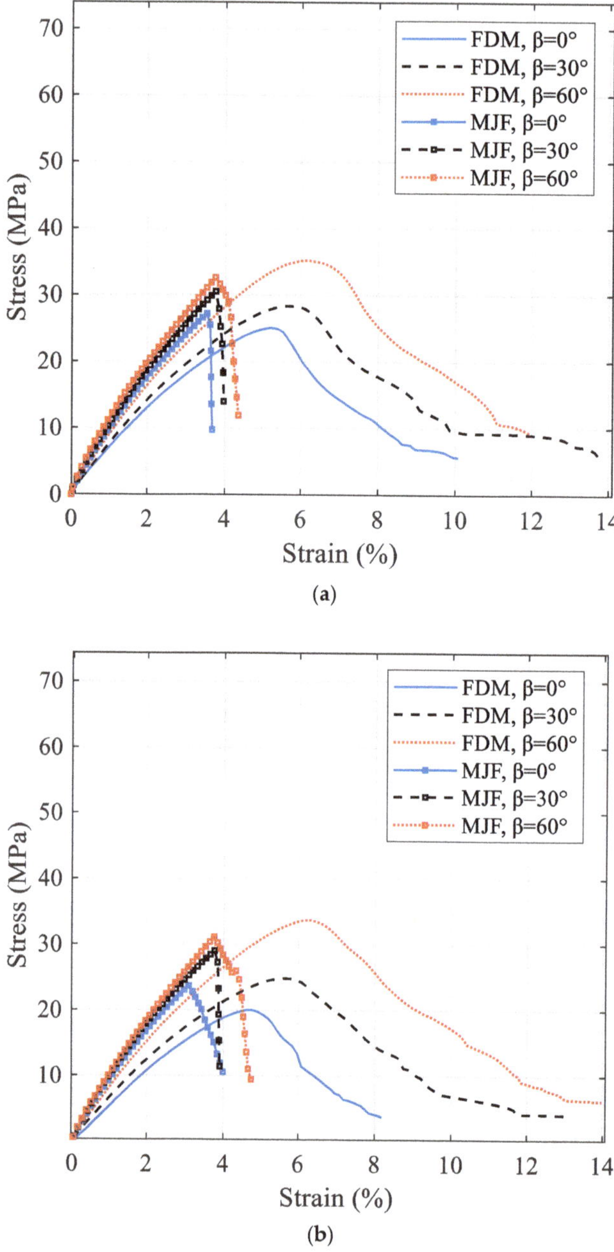

Figure 6. Standard tensile stress–strain curves for the FDM and MJF 3D-printed nylon; (**a**) $r = 1$ mm, (**b**) $r = 2$ mm

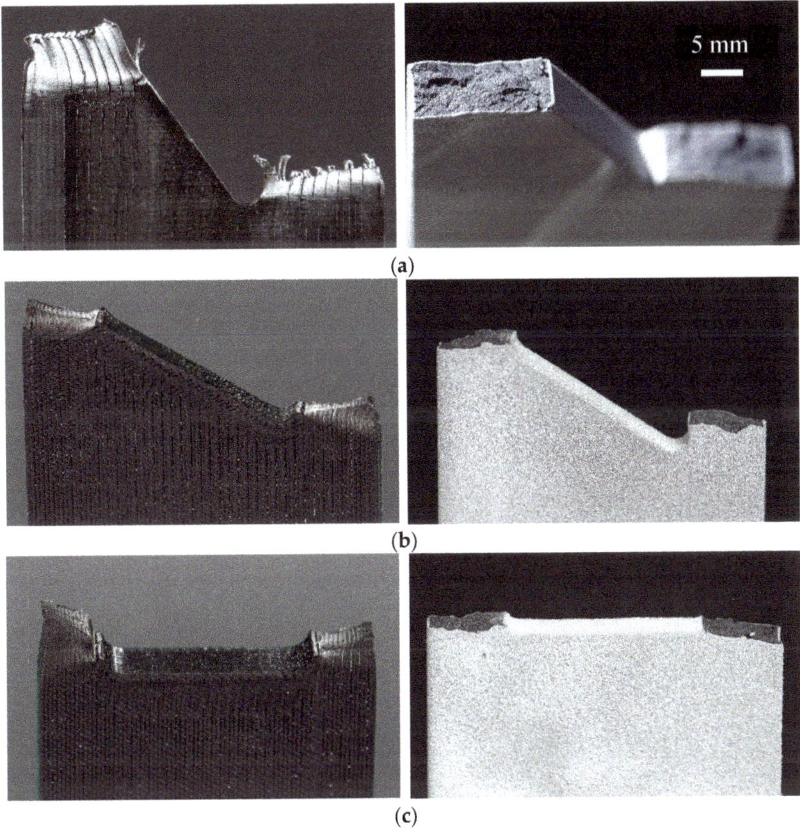

Figure 7. Close-up photographs FDM and MJF printed nylon specimens after fracture from left to right, respectively, with (**a**) 60°, (**b**) 30°, and (**c**) 0° notch orientations.

4. EMC, the J-Integral for U-Notched and THEIR Combination

Since connecting two or more parts is necessary in various specific designs, numerous engineering components contain notches with different shapes and sizes. Although existence of cracks is usually undesirable in engineering components, notches with different shapes (e.g., O, V, and U) are often unavoidable as parts of certain design requirements. However, notches have a distinct disadvantage that is stress concentrations in their vicinity that can lead to crack initiation, and ultimately failure of the component. In this section, we describe the theory of the EMC and then briefly review the J-integral in U-notched specimens, and finally explain their combination.

4.1. The Equivalent Material Concept (EMC)

Due to the complex and time-consuming nature of the analysis of the failure of notched ductile materials by crack propagation the equivalent material concept (EMC) was introduced [22]. This novel theory was proposed for predicting load-carrying capacity of notched components made of ductile materials. In EMC, a ductile material with elastic-plastic behavior and a valid K-based fracture toughness (K_C or K_{IC}) is considered to be equal to a virtual brittle material with an ideal linear elastic behavior to the breaking point. In this mode, the virtual material is assumed to have the same Poisson's ratio, Young's modulus and K-based fracture toughness as the real ductile material, but different tensile strength. The fracture toughness can be determined either by a direct approach, where pre-notched specimens are impact tested, or an indirect method based on the examination of the notched samples [23]. In EMC a ductile material is assumed to be equal to a virtual brittle material (with linear elastic behavior), hence we can employ brittle failure criteria in linear elastic fracture mechanics in order to predict ductile failure of notched components. According to this assumption, both materials absorb the same amount of strain energy density (SED) for the crack initiation, and the tensile strength of the equivalent material calculated.

Then the tensile strength and the fracture toughness can be simultaneously utilized in brittle fracture criteria by linear elastic analysis to predict the load-carrying capacity of notched ductile parts. As the SED absorbed by the ductile material is equal to a linear elastic SED absorbed by equivalent material, the ultimate tensile strength of equivalent material would be significantly larger than of the real ductile material. In other words, equating the two energy values would result in a larger ultimate strength in the linear stress versus strain relationship. The parameters E, σ_Y, ε_Y, ε_c, σ_u^* and denote elastic modulus, yield stress, yield strain, strain at crack initiation and tensile strength of the equivalent material respectively. It should be noted that the final rupture would be due to the brittleness of the material.

The power-law expression for the true stress–strain relationship in the plastic zone of the ductile material is as follows:

$$\sigma = K\varepsilon_p^n \tag{1}$$

where σ and K are the true stress and strain hardening coefficient, respectively. Also, ε_p and n denote the true plastic strain and the strain-hardening exponent, respectively. The total strain energy density consists of elastic and plastic components and can be written as follows:

$$(SED)_{total} = (SED)_{elastic} + (SED)_{plastic} = \frac{1}{2}\sigma_Y\,\varepsilon_Y + \int_{\varepsilon_y}^{\varepsilon_p} \sigma_p\,d\varepsilon_p \tag{2}$$

As $\varepsilon_Y = \sigma_Y/E$, with substitution of Equation (1) into Equation (2), results:

$$(SED)_{total} = \frac{\sigma_Y^2}{2E} + \int_{\varepsilon_Y}^{\varepsilon_p} K\varepsilon_p^n\,d\varepsilon_p \tag{3}$$

By integration, Equation (3) results in

$$(SED)_{total} = \frac{\sigma_Y^2}{2E} + \frac{K}{n+1}\left[\left(\varepsilon_p^{n+1} - \varepsilon_Y^{n+1}\right)\right] \tag{4}$$

Offset yield stress σ_Y is determined from a strain value of 0.2%.

$$(SED)_{total} = \frac{\sigma_Y^2}{2E} + \frac{K}{n+1}\left(\varepsilon_p^{n+1} - \varepsilon_Y^{n+1}\right) \tag{5}$$

where ε_Y is the yield strain.

The total strain energy density at any point in the plastic region can be calculated by Equation (5). By replacing ε_p with ε_c, where ε_c is the value of the strain at crack initiation, the total strain energy density associated with the onset of crack initiation can be determined:

$$(SED)_{total} = \frac{\sigma_Y^2}{2E} + \frac{K}{n+1}\left(\varepsilon_c^{n+1} - \varepsilon_Y^{n+1}\right) \tag{6}$$

Considering Equation (1), ε_c is determined as follows:

$$\varepsilon_c = \left(\frac{\sigma_c}{K}\right)^{1/n} \tag{7}$$

By substituting Equation (7) into Equation (6), we have

$$(SED)_{total} = \frac{\sigma_Y^2}{2E} + \frac{K}{n+1}\left(\left(\frac{\sigma_c}{k}\right)^{n+1/n} - (\varepsilon_Y)^{n+1}\right) \tag{8}$$

In EMC, the equivalent material is considered to be a virtual brittle material with the same value of elastic modulus and plane-strain fracture, but unknown value of ultimate tensile strength. The SED for this material at the onset of crack initiation is equal to:

$$(SED)_{EMC} = \frac{\varepsilon_u^{**2}}{2E} \tag{9}$$

As it was mentioned earlier, in EMC, it is assumed that SED values of real ductile and the virtual brittle material are equal. Hence, Equations (8) and (9) are identical. Therefore

$$\frac{\varepsilon_u^{**2}}{2E} = \frac{\sigma_Y^2}{2E} + \frac{K}{n+1}\left(\left(\frac{\sigma_c}{k}\right)^{n+1/n} - (\varepsilon_Y)^{n+1}\right) \tag{10}$$

Finally, the tensile strength σ_u^* of the equivalent material can be extracted as follows:

$$\sigma_u^* = \sqrt{\sigma_Y^2 + \frac{2EK}{n+1}\left(\left(\frac{\sigma_c}{k}\right)^{n+1/n} - (\varepsilon_Y)^{n+1}\right)} \tag{11}$$

Material fracture toughness and the tensile strength σ_u^* of the equivalent material are recognized as two necessary inputs in various brittle fracture criteria, to predict the crack initiation from the notch in ductile components under tensile loading regime.

4.2. A Brief Review of J-Integral in U-Notches

Existence of cracks in materials and their subsequent propagation under applied stresses often lead to unexpected failures, causing reliability issues in engineering structures and uncertainties over expected service life. Understanding the influence of notches and cracks of various sizes, shapes, with respect to applied stresses and material properties are important to predict the reliability of engineering components. To address this, failure criteria have been developed to predict the load-carrying capacities of engineering materials [24–26]. For ductile materials under general yielding condition energy dissipation [27] or J-integral criteria [28] are common powerful methods of analysis of crack propagation. In J-integral method a contour integral is defined which calculates the strain energy release rate needed to create two new surfaces in the cracked component under loading. Subsequently, some modifications of the J-integral concept were applied to different loading conditions [29–31]. The conditions for the path independence of the J-integral in U-and V-notched specimens [32], an analytical calculation of the J-integral for a sample with a crack initiated from a notch under elastic-plastic loading [33], and an expansion of J-integral concept considering volume element at the border of a sharp V-notch [34] were investigated.

Here, we analyze 3D-printed U-notched rectangular specimens for ductile failure under mix mode I/II loading by means of J-integral criterion, which is described considering basic equations of this fracture parameter. As proposed [28], absolute value of the J-integral for the cracked components is as follows:

$$J_k = \int_{\varphi}\left(Wn_k - T_i\frac{\partial u_i}{\partial x_k}\,ds\right), \quad (k = 1, 2) \tag{12}$$

where n_k is the unit vector normal to the specified contour path φ, u_i and T_i denote displacement and traction vectors. W and J_k are the strain energy density and the value of the J-integral, respectively. The value of the J-integral can be divided to two distinct values along specified axes. We assume x axis to be parallel to the notch bisector line, and located at the notch center of curvature (Figure 8).

The values of two J-integral parameters J_1 and J_2 can be expressed along the coordinate axes x and y as follows:

$$J_1 = \int_\varphi \left(Wdy - T_i \frac{\partial u_i}{\partial x} \, ds \right) \tag{13}$$

$$J_2 = \int_\varphi \left(Wdx - T_i \frac{\partial u_i}{\partial y} \, ds \right) \tag{14}$$

In the case of mixed mode I/II loading scenario, both Equations (13) and (14) have specific non-zero values and the equivalent J-integral value $\left(J_{eq} \right)$ can be determined by:

$$J_{eq} = \sqrt{J_1^2 + J_2^2} \tag{15}$$

In calculation of the J-integral for un-notched specimen under mixed mode I/II loading conditions, the inner section of the indicated control volume, which incorporates the notch border must be considered. Berto et al. [35] determined the specific control volume for U-notches to have a crescent shape. This is illustrated in Figure 7, where position of the control volume for U- notch under mode I and mixed mode I/II can be seen. The critical radius of curvature of the notch R_c, which is dependent on material properties and the plane-strain condition, expressed as follows:

$$R_c = \frac{(1+v)(5-8v)}{4\pi} \left(\frac{K_{Ic}}{\sigma_u} \right)^2 \tag{16}$$

where v, K_{Ic} and σ_u are Poisson's ratio, the plane-strain fracture toughness, and tensile strength, respectively. It should be pointed out that the thickness of the 3D-printed notched specimens are equal to 4 mm, which results in a significant plastically deformed zone around the notch at fracture. Because of the finite thickness and the resulting plastic zone, fracture toughness (K_c) was considered and inserted in Table 1 instead of the plane-strain fracture toughness (K_{Ic}). The use of K_c is appropriate as it is not a material property and depends on the sample thickness and relevant to large-scale plastic deformation occurring in the 3D-printed notched nylon specimens.

As mentioned earlier, equation for the critical radius of curvature of the notch is based on plane strain conditions in brittle failure. In order to apply it to our case K_{Ic} was substituted by K_c to predict ductile failure of the U-notched 3D-printed plastic specimens.

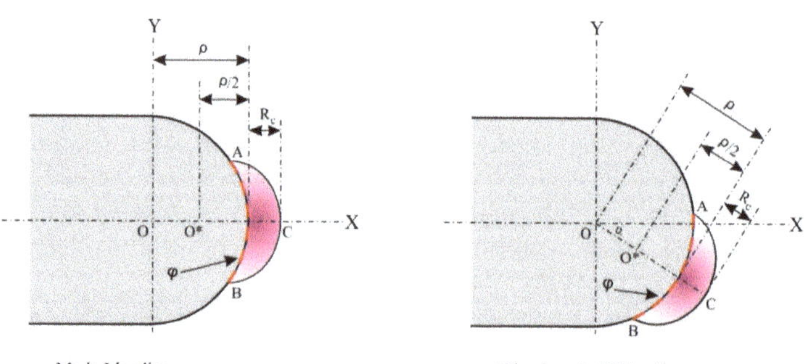

Figure 8. Material dependent control volume for a U-notch under mode I (**left**), and mixed mode I/II (**right**) loadings.

As it is illustrated in Figure 8, the outer radius of the crescent-shaped control volume is equal to $\frac{\rho}{2} + R_c$. The path ACB is considered as the contour path for J-integral calculation. Due to the traction-free surface on the contour path ACB, the second term of J-integral equations are equal to zero (Equations (13) and (14)).

4.3. Combination of EMC and J-Integral

In this section, a new expression of the J-integral criterion is described which combines the equivalent material concept and the J-integral in order to predict load-carrying capacity of the U-notched 3D-printed specimens. J-integral criterion is extended by using the material properties obtained from EMC to analyze the fracture in ductile U-notched samples. According to the J-integral criterion, brittle failure occurs when the value of the J-integral over a contour path along the inner arc reaches the critical value of J-integral (J_{cr}), which is linked to the critical strain energy density W_{cr} through the inner arc ACB in Figure 8 as:

$$J_{cr} \cong W_{cr} \times \text{arc} \, (ACB) \tag{17}$$

W_{cr} can be evaluated according to

$$W_{cr} = \frac{(\sigma_u)^2}{2E} \tag{18}$$

where σ_u is the tensile strength. By considering the critical radius and critical strain energy density of the equivalent material J-integral criterion can be used for the prediction of the ductile fracture in U-notched specimens. By using the tensile strength σ_u^* of the equivalent material instead of σ_u in Equations (16) and (18) the critical radius of the R_{CE}, and the critical strain energy density W_{CE} for the equivalent material can be determined as follows:

$$R_{CE} = \frac{(1+v)(5-8v)}{4\pi} \left(\frac{K_c}{\sigma_u^*} \right)^2 \tag{19}$$

$$W_{CE} = \frac{(\sigma_u^*)^2}{2E} \tag{20}$$

Substituting values obtained by Equation (20) into Equation (17):

$$J_{CE} = \frac{(\sigma_u^*)^2}{2E} \times \text{arc} \, (ACB) \tag{21}$$

where, J_{CE} is the value of the J-integral for the tensile tested U-notched 3D-printed MJF and FDM nylon samples by utilizing the combined EMC/J-integral criterion. To achieve this, the strain energy density value along the specified contour path is needed, which can be evaluated by numerical simulation in finite element software.

5. Numerical Simulations and Results

With the aim to conform the experimental finding, we simulated the U-notched specimens that experienced tensile loads, and finite element analysis was performed. In this context, ABAQUS software was used to analyze the strain energy density distribution along the contour path defined in the vicinity of the notch. For this purpose, a finite element model was created for each U-notched specimen, utilizing eight-node biquadratic plane-strain quadrilateral element types. Considering the thickness of the specimens, they have been considered slender. In simulation, the meshes have been refined at the notch tip vicinity, because of high level of stress gradient. In numerical simulations, the tensile load was applied to the nodes that lie on the upper end of the modeled U-notched specimens. The nodes, which are lied on the lower end of the rectangular model, were constrained to be completely fixed.

The strain energy density contours around the notch for three selected specimens under tensile loading are illustrated in Figure 9. The parameters J_1, J_2, J_{eq}, arc (ACB), and J_{CE} are presented in Table 3. It should be noted that these values are obtained for the average critical load (P_{av}) in the tensile experiments. It should be noted that the obtained tensile stress versus strain curve was utilized in this analysis. Considering different notch radii, there are different control volume geometries for the model. Based on the method described in the previous section, the control volume is centered along the bisector line of the notch. In addition, by considering the value of *RCE*, the value of arc (*ACB*) is obtained from FEA.

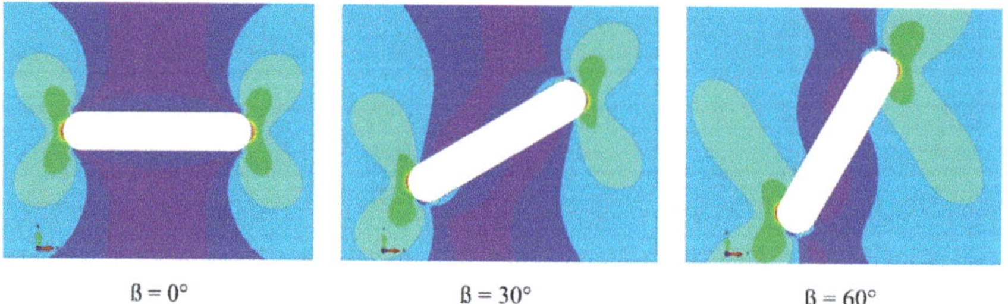

Figure 9. Illustration of strain energy density around the notch border for MJF nylon with a notch tip radius of 2 mm at different notch orientations.

Table 3. Numerical results of 3D-printed U-notched nylon material under mode I and mixed mode I/II loading scenarios.

Material	β (deg.)	ρ (mm)	P_{av} (N)	J_1 (N/mm)	J_2 (N/mm)	J_{eq} (N/mm)	arc (ACB) (mm)	J_{CE} (N/mm)
FDM Nylon	0	1	5011.8	12.7	0	12.7	1.3	7.3
		2	4015.1	12.1	0	12.1	1.1	6.1
	30	1	5667.3	11.8	11.2	16.2	1.6	8.4
		2	4968.6	10.7	9.5	14.3	1.2	6.2
	60	1	7047.5	90.3	9.4	13.2	1.7	9.5
		2	6714.7	8.5	9.1	12.4	1.4	8.4
MJF Nylon	0	1	5441.2	12.9	0	12.9	1.2	5.3
		2	4749.6	12.3	0	12.3	1.1	4.8
	30	1	5547.7	11.7	11.3	16.2	1.3	5.7
		2	5802.8	10.9	9.8	14.6	1.1	4.8
	60	1	6520.2	8.2	8.7	11.9	1.4	6.2
		2	6219.2	7.4	8.1	10.9	1.2	5.3

A ductile failure model, called EMC-J criterion to evaluate load-carrying capacity of a U-notched 3D-printed ductile polymer samples under mixed mode I/II loading is introduced. At first by utilizing Equation (15) the value of equivalent J-integral for an assumed load (e.g., 1 N) is determined, then the load increased till J_{eq} attains J_{CE}. The critical load can therefore be evaluated by the following equation:

$$\frac{P_{cr}}{P_{applied}} = \sqrt{\frac{J_{CE}}{J_{eq}}} \tag{22}$$

It should be pointed out that in EMC-J criterion, we considered a specified contour path round the notch border, and then theoretical results are evaluated by calculation of definite integral of SED values along this path.

In the current research, EMC-J criterion is employed to predict load-carrying capacity of U-notched 3D-printed materials from a series of experiments, which were performed on notch tip radii ($\rho = 1$ and 2 mm) with three notch inclination angles ($\beta = 0°$, 30°, 60°). In Figure 10, average critical load is expressed as a function of the loading angle (β) and experimental findings are compared with results achieved by the EMC-J criterion. Results indicate that the critical load increases with the loading angle in each notch tip radius. As expected, the contribution of mode II loading increased with the increase of loading angle. That is, increase in the notch inclination angle β led to a greater plastically deformed zone around the notch at failure, manifesting as greater resistance to the applied load. As seen in Figure 10 and Table 4, the experimental findings are in good agreement with the theoretical predictions realized from EMC-J-integral criterion.

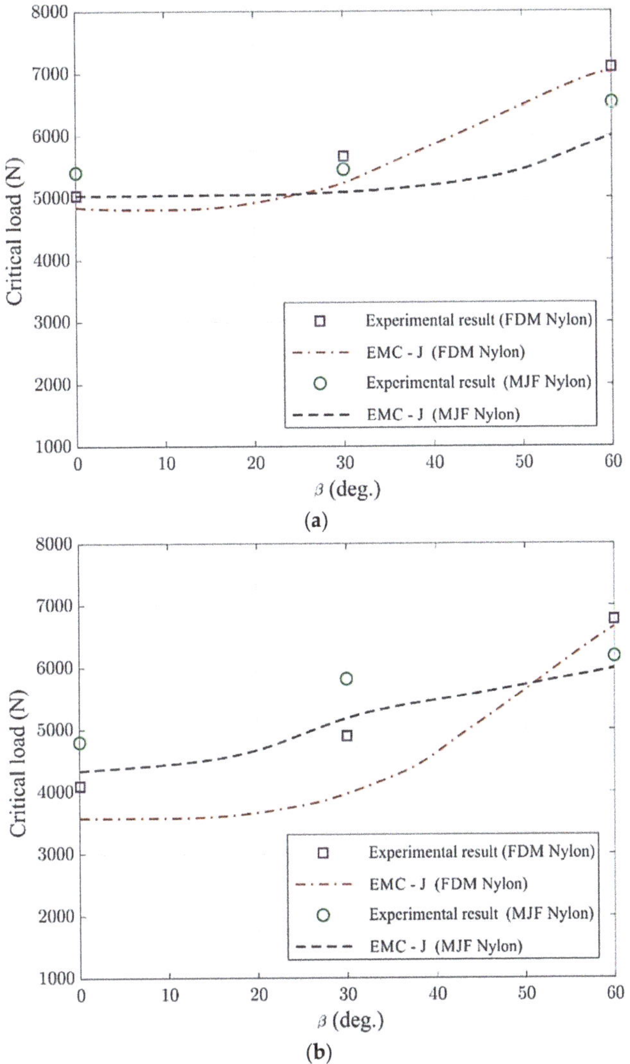

Figure 10. 3D-printed nylon specimens U-notch comparisons via EMC-J criterion; (**a**) $\rho = 1$ mm, (**b**) $\rho = 2$ mm.

The experimental results of load-carrying capacity with theoretical predictions achieved from EMC-J criterion for the examined 3D-printed nylon specimens and differences between the experimental findings and theoretical results are presented in Table 4. There seems to a difference of up to 20%, between experimental findings and the results achieved from EMC-J, but the difference in most of the cases is lower than 8%. Therefore these findings suggest that EMC-J-integral criterion can be considered as a reliable approach to predict ductile failure of U-notched 3D-printed nylon components subjected to mixed mode I/II loading conditions.

Table 4. Calculated critical load for studied 3D-printed specimens by means of EMC-J criterion.

Material	β (deg.)	ρ (mm)	P_{av} (N)	P_{EMC-J} (N)	Δ_{EMC-J} (%)
FDM Nylon	0	1	5011.8	4726.1	5.7
		2	4015.1	3583.4	10.7
	30	1	5667.3	5218.6	7.9
		2	4968.6	3995.2	19.5
	60	1	7047.5	7003.2	0.6
		2	6714.7	6581.9	1.9
MJF Nylon	0	1	5441.2	5022.3	7.7
		2	4749.6	4362.1	8.1
	30	1	5547.7	5042.5	9.1
		2	5802.8	5245.7	9.6
	60	1	6520.2	5973.4	8.3
		2	6219.2	5912.2	4.9

6. Conclusions

There has been recent growth in the additive manufacturing of engineering components, in addition to the number of various three-dimensional (3D) printing techniques. However, the parts fabricated by different techniques differ in terms of strength, stiffness, microstructure and material properties. Therefore, investigation of fracture behaviour is a vital component of the engineering design. In this study, we presented an analysis of the fracture of 3D-printed thermoplastic components made by fused deposition modeling (FDM) and multi-jet fusion (MJF) 3D printing techniques. Nylon 12 filament and PA12 nylon powder were used for 3D printing specimens with centrally positioned notches. It was observed that the printing techniques employed resulted in pronounced differences in tensile behavior of the 3D-printed components.

In the MJF method, a uniform heat is applied to fuse the powder into an isotropic solid layer, whereas in the case of the FDM the thermoplastic filament is melted as it is positioned to form each layer. The continuous nature of the filament aligned in the direction of the applied stress have resulted in better structural integrity in the FDM samples than that achieved by MJF, thus resulting in significantly higher elongation at break and toughness. However, the mechanical properties of an FDM 3D-printed part is far more complex than it may appear at first glance. It is well established that the FDM specimens are anisotropic with the greater tensile strength in the axial direction than in the transverse direction normal to the bonds [7].

Irrespective of the notch orientations, all MJF samples exhibited brittle behavior with flat fracture surfaces. Considering the tensile tests, the failure load of MJF 3D-printed nylon was observed to be greater than the FDM samples for $\beta = 0°$ and $30°$, while this was not the case when the inclination angle increased to $60°$. In addition to the effect of crack angles, it was observed that increase in crack radius was associated with reduced critical load in both types of 3D-printed samples.

Finally, the equivalent material concept (EMC) was combined with the J-integral failure principle to predict the fracture failure of U-notched 3D-printed specimens subjected to tensile loading under mode I and mixed mode I/II loading regimes. The agreement between the experimental and simulation results proved the EMC-J approach to be capable of successfully predicting fracture in the 3D-printed notched ductile material components.

Author Contributions: Conceptualization, M.R.K.; Investigation, A.Z.; Methodology, M.R.K.; Validation, A.K.; Resources, A.Z.; Writing-Original Draft Preparation, M.R.K. and A.Z.; Writing-Review & Editing, A.K.; Visualization, A.K. and M.R.K.; Supervision, A.K.; Project Administration, A.Z. and M.R.K.; Funding Acquisition, A.Z. All authors have read and agreed to the published version of the manuscript.

Funding: This research received no external funding.

Conflicts of Interest: The authors declare no conflict of interest.

References

1. Libonati, F.; Gu, G.X.; Qin, Z.; Vergani, L.; Buehler, M.J. Bone-Inspired Materials by Design: Toughness Amplification Observed Using 3D Printing and Testing. *Adv. Eng. Mater.* **2016**, *18*, 1354–1363. [CrossRef]
2. Papon, E.A.; Haque, A. Fracture toughness of additively manufactured carbon fiber reinforced composites. *Addit. Manuf.* **2019**, *26*, 41–52. [CrossRef]
3. Zolfagharian, A.; Kaynak, A.; Kouzani, A. Closed-loop 4D-printed soft robots. *Mater. Des.* **2020**, *188*. [CrossRef]
4. Mitsouras, D.; Liacouras, P.C. 3D printing technologies. In *3D Printing in Medicine*; Springer: Berlin, Germany, 2017; pp. 5–22.
5. Zolfagharian, A.; Kouzani, A.Z.; Khoo, S.Y.; Moghadam, A.A.A.; Gibson, I.; Kaynak, A. Evolution of 3D printed soft actuators. *Sens. Actuators A Phys.* **2016**, *250*, 258–272. [CrossRef]
6. Zhou, Y.; Zou, J.; Wu, H.; Xu, B. Balance between bonding and deposition during fused deposition modeling of polycarbonate and acrylonitrile-butadiene-styrene composites. *Polym. Compos.* **2019**, *41*, 60–72. [CrossRef]
7. Turner, B.N.; Strong, R.; Gold, S.A. A review of melt extrusion additive manufacturing processes: I. Process design and modeling. *Rapid Prototyp. J.* **2014**, *20*, 192–204. [CrossRef]
8. Thrimurthulu, K.; Pandey, P.M.; Reddy, N.V. Optimum part deposition orientation in fused deposition modeling. *Int. J. Mach. Tools Manuf.* **2004**, *44*, 585–594. [CrossRef]
9. Zhou, Y.-G.; Su, B.; Turng, L.-S. Deposition-induced effects of isotactic polypropylene and polycarbonate composites during fused deposition modeling. *Rapid Prototyp. J.* **2017**, *23*, 869–880. [CrossRef]
10. Zolfagharian, A.; Kouzani, A.Z.; Khoo, S.Y.; Gibson, I.; Kaynak, A. 3D printed hydrogel soft actuators. In Proceedings of the 2016 IEEE Region 10 Conference (TENCON), Singapore, 22–25 November 2016.
11. Gardan, J.; Makke, A.; Recho, N. Improving the fracture toughness of 3D printed thermoplastic polymers by fused deposition modeling. *Int. J. Fract.* **2018**, *210*, 1–15. [CrossRef]
12. Li, T.; Chen, Y.; Wang, L. Enhanced fracture toughness in architected interpenetrating phase composites by 3D printing. *Compos. Sci. Technol.* **2018**, *167*, 251–259. [CrossRef]
13. Khosravani, M.R. Influences of Defects on the Performance of Adhesively Bonded Sandwich Joints. *Key Eng. Mater.* **2018**, *789*, 45–50. [CrossRef]
14. Mirzaali, M.J.; Edens, M.E.; De La Nava, A.H.; Janbaz, S.; Vena, P.; Doubrovski, E.L.; Zadpoor, A.A. Length-scale dependency of biomimetic hard-soft composites. *Sci. Rep.* **2018**, *8*, 12052. [CrossRef] [PubMed]
15. Ayatollahi, M.; Torabi, A. Determination of mode II fracture toughness for U-shaped notches using Brazilian disc specimen. *Int. J. Solids Struct.* **2010**, *47*, 454–465. [CrossRef]
16. Luo, Y.; Xie, H.P.; Ren, L.; Zhang, R.; Li, C.B.; Gao, C. Linear Elastic Fracture Mechanics Characterization of an Anisotropic Shale. *Sci. Rep.* **2018**, *8*, 8505. [CrossRef]
17. Carpinteri, A.; Cornetti, P.; Pugno, N.; Sapora, A. On the most dangerous V-notch. *Int. J. Solids Struct.* **2010**, *47*, 887–893. [CrossRef]
18. Sapora, A.; Cornetti, P.; Carpinteri, A. A Finite Fracture Mechanics approach to V-notched elements subjected to mixed-mode loading. *Eng. Fract. Mech.* **2013**, *97*, 216–226. [CrossRef]
19. Pippan, R.; Hohenwarter, A. The importance of fracture toughness in ultrafine and nanocrystalline bulk materials. *Mater. Res. Lett.* **2016**, *4*, 127–136. [CrossRef]
20. Gómez, F.; Torabi, A. Application of the equivalent material concept to the study of the ductile failure due to U-notches. *Int. J. Press. Vessel. Pip.* **2019**, *172*, 65–69. [CrossRef]
21. Majidi, H.; Golmakani, M.; Torabi, A. On combination of the equivalent material concept and J-integral criterion for ductile failure prediction of U-notches subjected to tension. *Fatigue Fract. Eng. Mater. Struct.* **2018**, *41*, 1476–1487. [CrossRef]
22. Torabi, A. Estimation of tensile load-bearing capacity of ductile metallic materials weakened by a V-notch: The equivalent material concept. *Mater. Sci. Eng. A* **2012**, *536*, 249–255. [CrossRef]
23. Berto, F.; Gomez, G. Notched plates in mixed mode loading (I+ II): A review based on the local strain energy density and the cohesive zone mode. *Eng. Solid Mech.* **2017**, *5*, 1–8. [CrossRef]
24. Lazzarin, P.; Zambardi, R. A finite-volume-energy based approach to predict the static and fatigue behavior of components with sharp V-shaped notches. *Int. J. Fract.* **2001**, *112*, 275–298. [CrossRef]
25. Khosravani, M.R.; Anders, D.; Weinberg, K. Influence of strain rate on fracture behavior of sandwich composite T-joints. *Eur. J. Mech. A/Solids* **2019**, *78*. [CrossRef]

26. Carloni, C.; Nobile, L. Maximum circumferential stress criterion applied to orthotropic materials. *Fatigue Fract. Eng. Mater. Struct.* **2005**, *28*, 825–833. [CrossRef]

27. Turner, C.E.; Kolednie, O. A Micro and Macro Approach to the Energy Dissipation Rate Model of Stable Ductile Crack Growth. *Fatigue Fract. Eng. Mater. Struct.* **1994**, *17*, 1089–1107. [CrossRef]

28. Rice, J.R. Stresses Due to a Sharp Notch in a Work-Hardening Elastic-Plastic Material Loaded by Longitudinal Shear. *J. Appl. Mech.* **1967**, *34*, 287–298. [CrossRef]

29. Frank, D.; Remes, H.; Romanoff, J. J-integral-based approach to fatigue assessment of laser stake-welded T-joints. *Int. J. Fatigue* **2013**, *47*, 340–350. [CrossRef]

30. Yoshihara, H.; Yoshinobu, M. Determination of the Mode I critical J-integral of copy paper. *Eng. Fract. Mech.* **2017**, *169*, 251–261. [CrossRef]

31. Meneghetti, G.; Ricotta, M.; Pitarresi, G. Infrared thermography-based evaluation of the elastic-plastic J-integral to correlate fatigue crack growth data of a stainless steel. *Int. J. Fatigue* **2019**, *125*, 149–160. [CrossRef]

32. Chen, Y.-H.; Lu, T.J. On the path dependence of the J-integral in notch problems. *Int. J. Solids Struct.* **2004**, *41*, 607–618. [CrossRef]

33. Matvienko, Y.; Morozov, E. Calculation of the energy J-integral for bodies with notches and cracks. *Int. J. Fract.* **2004**, *125*, 249–261. [CrossRef]

34. Berto, F.; Lazzarin, P. Relationships between J-integral and the strain energy evaluated in a finite volume surrounding the tip of sharp and blunt V-notches. *Int. J. Solids Struct.* **2007**, *44*, 4621–4645. [CrossRef]

35. Berto, F.; Lazzarin, P.; Marangon, C. Brittle fracture of U-notched graphite plates under mixed mode loading. *Mater. Des.* **2012**, *41*, 421–432. [CrossRef]

Article

In-Plane Mechanical Behavior of a New Star-Re-Entrant Hierarchical Metamaterial

Wenjiao Zhang [1,*]**, Shuyuan Zhao** [2]**, Rujie Sun** [3]**, Fabrizio Scarpa** [3] **and Jinwu Wang** [1,*]

[1] School of Engineering, Northeast Agricultural University, No. 600 Changjiang Road, Harbin 150030, China
[2] Center for Composite Materials and Structures, Harbin Institute of Technology, Harbin 150080, China
[3] Bristol Composites Institute (ACCIS), University of Bristol, Bristol BS8 1TR, UK
* Correspondence: zhangwenjiao@neau.edu.cn (W.Z.); jinwuw@163.com (J.W.)

Received: 11 May 2019; Accepted: 21 June 2019; Published: 3 July 2019

Abstract: A novel hierarchical metamaterial with tunable negative Poisson's ratio is designed by a re-entrant representative unit cell (RUC), which consists of star-shaped subordinate cells. The in-plane mechanical behaviors of star-re-entrant hierarchical metamaterial are studied thoroughly by finite element method, non-dimensional effective moduli and effective Poisson's ratios (PR) are obtained, then parameters of cell length, inclined angle, thickness for star subordinate cell as well as the amount of subordinate cell along x, y directions for re-entrant RUC are applied as adjustable design variables to explore structure-property relations. Finally, the effects of the design parameters on mechanical behavior and relative density are systematically investigated, which indicate that high specific stiffness and large auxetic deformation can be remarkably enhanced and manipulated through combining parameters of both subordinate cell and parent RUC. It is believed that the new hierarchical metamaterial reported here will provide more opportunities to design multifunctional lightweight materials that are promising for various engineering applications.

Keywords: hierarchical; metamaterial; re-entrant structure; auxetic; mechanical behavior

1. Introduction

Re-entrant honeycomb structures that display negative Poisson's ratios (NPR) are known to be one class of auxetic structures and have been used in many fields, such as aerospace and automotive industries. The multifunctionality of anisotropic re-entrant honeycomb has been widely studied for its static mechanical behavior [1–4], dynamic performance [5,6], thermal conductivity and heat transfer properties [7].

Hierarchy [8] is one of the most readily observed topological features in natural structures and now has been introduced to honeycomb and chiral lattice structures in pursuing ultralight materials with improving elastic properties and damage tolerance. Specifically, considering hierarchical sub-structures to honeycombs and designing novel metamaterials with tailorable multi-functional properties, have attracted increasing attention in recent years.

Extreme values of hierarchical metamaterial properties such as specific stiffness, toughness, strength, negative or complex Poisson's ratio, zero or negative thermal expansion, phononic band gaps as well as impact energy absorption have been reported in hierarchical architectures across multiple length scales [9–15]. Sun et al. [16] analytically studied the in-plane elastic moduli and thermal conductivity of a multifunctional hierarchical honeycomb (MHH), which is formed by replacing the solid cell walls of an original regular hexagonal honeycomb (ORHH) with three different isotropic honeycomb sub-structures possessing hexagonal, triangular or kagome lattices. Then the anisotropic multifunctional hierarchical honeycomb (AMHH) with triangular or kagome honeycomb substructures (OAHH) was proposed and the in-plane stiffness of these two kinds of AMHH was analytically studied with the help of Euler beam theory [17]. Taylor et al. [18] investigated the in-plane elastic properties and structural hierarchy in honeycombs and explored the effects of adding hierarchy into a range of honeycombs, with hexagonal, triangular or square geometry super and sub-structure cells by using finite element simulation. Key parameters describing these geometries included the relative lengths of the sub- and super-structures, the fraction of mass shared between the sub- and super-structures, the co-ordination number of the honeycomb cells, the form and extent of functional grading, and the Poisson's ratio of the sub-structure. Mousanezhad et al. [19] studied the effects of chirality and hierarchy on elastic response of honeycombs, derived the closed-form expressions for elastic moduli of several chiral, anti-chiral and hierarchical honeycombs with hexagon and square based networks, and finally validated the analytical estimates of the elastic moduli by using finite element method. Gatt et al. [20] proposed a new class of hierarchical auxetics based on the rotating rigid unit mechanism. These systems retain the enhanced properties from having a negative Poisson's ratio with the added benefits of being a hierarchical system. Through design, one can control the extent of auxeticity, degree of aperture and size of the different pores in the system, which makes the system more versatile than similar non-hierarchical ones. Chen et al. [21] reported a group of hierarchically architected metamaterials constructed by replacing cell walls of regular honeycombs with hexagonal, kagome, and triangular lattices, respectively. The numerical and analytical studies indicate that the introduction of structural hierarchy in regular honeycombs results in improved heat resistance and thermal anisotropy. Then, Yin et al. [22] studied the in-plane crashworthiness of the hierarchical honeycomb group above, using the nonlinear finite element code LS-DYNA. The numerical simulation results indicate that the triangular hierarchical honeycomb provides the best performance compared to the other two hierarchical honeycombs and features more than twice the energy absorbed by the regular honeycomb under similar loading conditions. More recently, Wu et al. [23,24] proposed an innovative hierarchical anti-tetrachiral structure as well as a hierarchical anti-tetrachiral stent with circular and elliptical nodes, based on the auxetic deformation behaviors of anti-tetrachiral unit cell at different structural hierarchical levels. It was found that the mechanical behaviors of hierarchical anti-tetrachiral structure can be tailored through adjusting the levels of hierarchical structures and unit cell design, and the proposed hierarchical anti-tetrachiral stents exhibit remarkable radial expanding abilities while maintaining axial stability. Besides the above-mentioned hierarchical metamaterials, various types of alternative hierarchical structures have been proposed through the modification of the node or cell wall structural levels for generating enhanced and tunable mechanical properties through structural hierarchy approaches.

In the current work, a novel auxetic hierarchical metamaterial was designed, which consisted of a re-entrant representative unit cell as well as a star subordinate cell with zero Poisson's ratio; both the main and sub cells were planar symmetric. Full dimensional models of the new hierarchical metamaterial to describe the effective elasticity as well as loading-bearing capability in plane were simulated by finite element method (FEM). Comprehensive parametric studies for both parent RUC and subordinate cell were performed to evaluate in-plane non-dimensional moduli, effective Poisson's ratio and relative density of the new hierarchical structure, corresponding optimum structure-property relations were explored for the designed metamaterial.

2. Materials and Methods

The geometry of the novel hierarchical star-re-entrant metamaterial is presented in Figure 1. The representative unit cell (RUC), composed of star subordinate cell can be described as a re-entrant hexagonal structure, and apparently both the cells are horizontal and vertical symmetry. Analogy to a zero Poisson's ratio cellular structure [25], the star subordinate cell is represented by length L_0, inclined angle θ as well as in-plane thickness t, respectively (Figure 1a). Then dimensions of the new hierarchical star-re-entrant RUC is described by length L_x, L_y and symmetric inclined angle 45° (Figure 1b), where $L_i = N_i \times L_0$ ($i = x, y$), N_i is the number of star subordinate cells along x or y direction. The detail of star joint in re-entrant structure is presented in Figure 1c, the length of jointing star is expressed as $W = (\sqrt{2}t)/(4\sin\theta\cos(45° - \theta))$. To avoiding the overlapping and contact of cell walls, geometric constraints $0° < \theta < 45°$ and $0 < t < L_0(\cos\theta - \sin\theta)$ need to be satisfied.

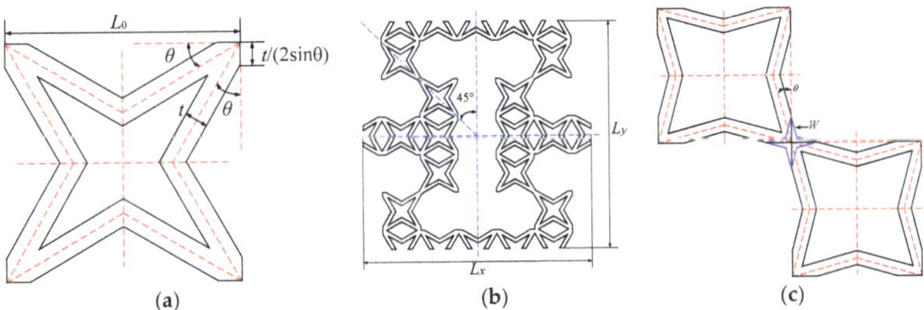

Figure 1. Geometry of the novel star-re-entrant hierarchical metamaterial: (**a**) star subordinate cell, (**b**) re-entrant representative unit cell, (**c**) the jointing between two neighboring stars.

The relative density is an important parameter for cellular structures, and it is defined by:

$$\frac{\rho}{\rho_c} = \frac{A}{A_c} \tag{1}$$

where A and A_c are respectively the cross-sections perpendicular to the out-plane thickness direction and the load bearing area. Here, for this new hierarchical RUC, relative density results in:

$$\frac{\rho}{\rho_c} = \frac{36L_0t\sin 3\theta - 36L_0t\sin\theta - 10t^2\sin 2\theta - 8t^2\cos 2\theta - 8t^2}{9L_0^2\sin 4\theta} \tag{2}$$

Finite element analysis was performed with Abaqus/CAE 6.13–4 commercial package standard for one RUC as well as the whole hierarchical metamaterial structure. For all the simulations in this paper, geometric dimensions of the new RUC are defined as $L_0 = 20$, $\theta = 30°$, $t = 2$ and $L_x = L_y = 6L_0$ with symmetric inclined angle 45°, respectively. Acrylonitrile butadiene styrene (ABS) plastic with a rapid prototyping Fusion Deposition Molding (FDM) Stratasys machine was used to manufacture all the experimental samples and the elastic mechanical properties of the core material for finite element simulation were set as $E_c = 2265$MPa and Poisson's ratio $v_c = 0.25$ [25,26]. An elastic shell element with reduced integration (S4R) and element size of 0.8 for convergence were chosen for the simulation of in-plane effective moduli, shown in Figure 2a. Accounting for the symmetry, the moduli of elasticity and the Poisson's ratio were determined by one quarter of the metamaterial structure. Taking full-size representative volumes with 6×6 cells for example, boundary conditions for in-plane tensile Young's moduli and Poisson's ratio were established based on References [23,27], where nodes on the left and bottom edge were constrained from out-plane rotation and translation normal to the edge direction, respectively, and displacements in the x-direction (y-direction) were applied to the ligament nodes on the right (top) edge, which was also constrained from in-plane rotation, shown in Figure 2b. In the case of the in-plane shear simulation, biaxial loading was introduced as close as possible a pure shear deformation field [27,28], corresponding boundary conditions above as well as displacements in both x and y directions were applied and are presented in Figure 2c.

Nominal strain and stress of the hierarchical representative unit cell in $i(= x$ or $y)$ direction were calculated from:

$$\varepsilon_i = \frac{\delta_i}{L_i}, \ \sigma_i = \frac{F_i}{A_j} \tag{3}$$

where δ_i is the applied displacement, F_i is the sum of the nodal reaction forces on the edge to which displacement was applied, L_i and A_j are the initial length and cross-sectional area of the hierarchical structure in the i and $j (= y$ or $x)$ directions, respectively. According to Equation (3), effective Young's modulus as well as Poisson's ratio are calculated by:

$$E_i = \frac{\sigma_i}{\varepsilon_i}, \quad \nu_{ij} = -\frac{\delta_j L_i}{\delta_i L_j} \tag{4}$$

where, i is the loading direction.

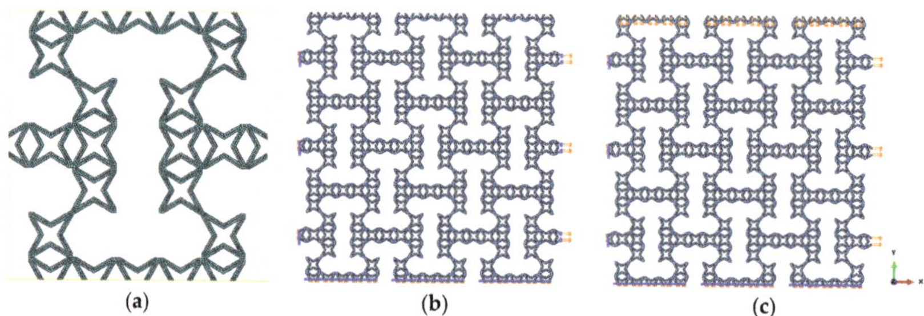

Figure 2. Numerical model description: (**a**) Mesh of one representative unit cell, (**b**) boundary conditions of axial tension along x direction, (**c**) boundary conditions of biaxial shear test.

Then, effective shear modulus under biaxial loading in plane was obtained as following [27,28]:

$$G_{xy} = \frac{\tau}{\gamma} = \frac{\sigma_x - \sigma_y}{2(\varepsilon_x - \varepsilon_y)} = \frac{R_x L_x - R_y L_y}{2h(\delta_x L_y - \delta_y L_x)} \tag{5}$$

In Equation (5), R_x and R_y are reaction force along x and y direction, h is the out-plane thickness of hierarchical structure.

In order to highlight the influence of the cell numbers on the convergence of the results, computations were undertaken starting by a number of 2 × 2 cells to a maximum of 40 × 40 cells. The convergence was found to be achieved at the number of 40 cells; corresponding dependence of mechanical property on the computations number of cells as well as the results of effective non-dimensional moduli and Poisson's ratio are demonstrated in Figure 3, respectively.

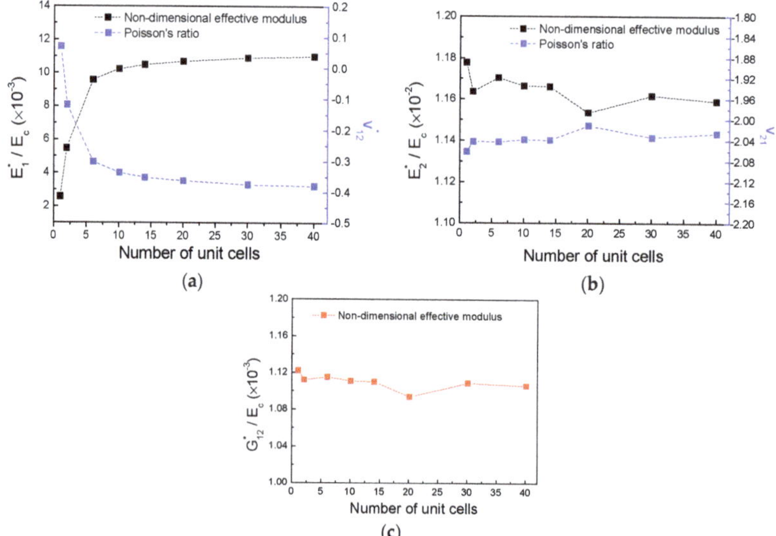

Figure 3. Finite element simulations of effective mechanical property for the new hierarchical metamaterial: (**a**) E_1^*/E_c and ν_{12}^*, (**b**) E_2^*/E_c and ν_{21}^*, (**c**) G_{12}^*/E_c.

3. Results and Discussion

To understand how the geometrical parameters of the star-re-entrant RUC influence the effective mechanical properties of the new hierarchical metamaterial designed, parametric studies were conducted by using finite element models described in Section 3, and the numerical results are presented and discussed as follows.

3.1. The Geometry Effects of Star Subordinate Cell

Figures 4–6 demonstrate the FE homogenization of the non-dimensional in-plane elastic moduli and corresponding Poisson's ratio versus various parameters of lengths L_0, thickness t and cell inclined angle θ. In general, all the non-dimensional elastic moduli decrease with increasing L_0, shown in Figure 4 and increase with increasing thickness t, seen in Figure 5, when the other geometrical parameters keep constant. The variations of non-dimensional effective moduli with cell inclined angle θ are presented in Figure 6. For the increase of θ, E_1^*/E_c exhibits an up-down-up trend, while the other two non-dimensional moduli display a first descent and then ascent with different gradients, for which E_2^*/E_c and G_{12}^*/E_c increased by 9.89% and 35.9%, respectively. The impact of geometric parameters on effective Poisson's ratio of the new hierarchical metamaterial are discussed as following. Poisson's ratio v_{12}^* only increases with increasing L_0 (Figure 4a) and declines with thickness t and cell inclined angle θ, presented in Figures 5a and 6a, respectively. While the variations of Poisson's ratio v_{21}^* with rising L_0, θ and t exhibit as constant (Figure 4b), up-down (Figure 5b) as well as ascending (Figure 6b), separately.

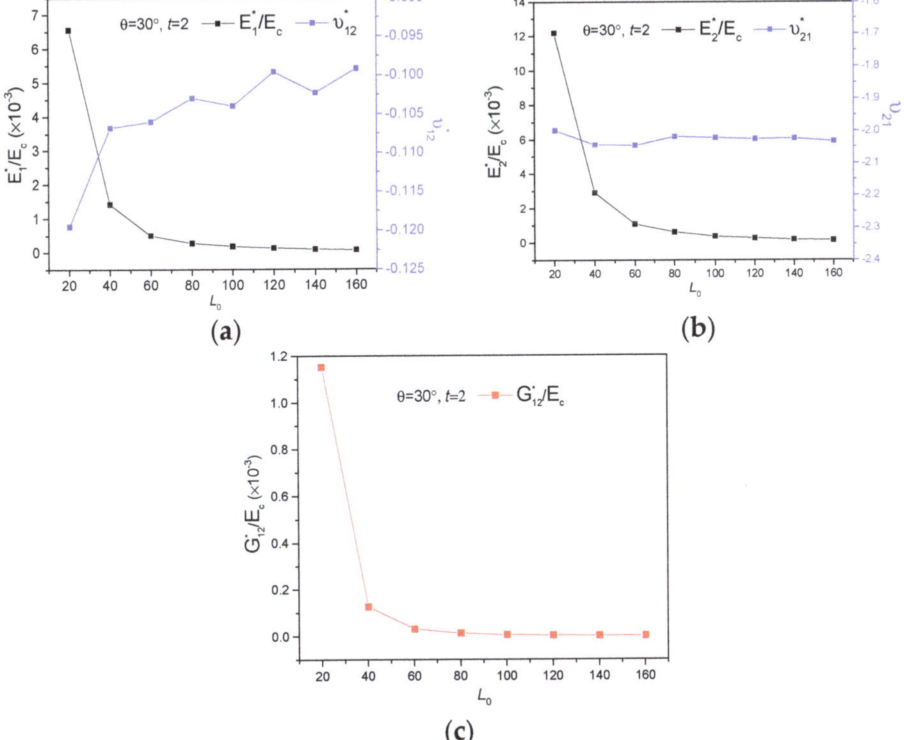

Figure 4. The variations of effective mechanical properties verse length L_0: (**a**) E_1^*/E_c and v_{12}^*, (**b**) E_2^*/E_c and v_{21}^*, (**c**) G_{12}^*/E_c.

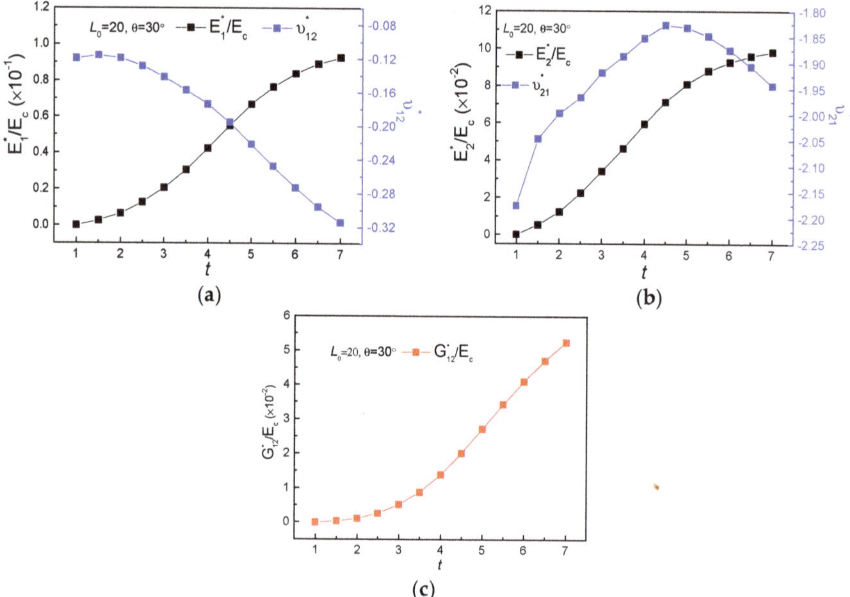

Figure 5. The variations of effective mechanical properties verse thickness t: (**a**) E_1^*/E_c and v_{12}^*, (**b**) E_2^*/E_c and v_{21}^*, (**c**) G_{12}^*/E_c.

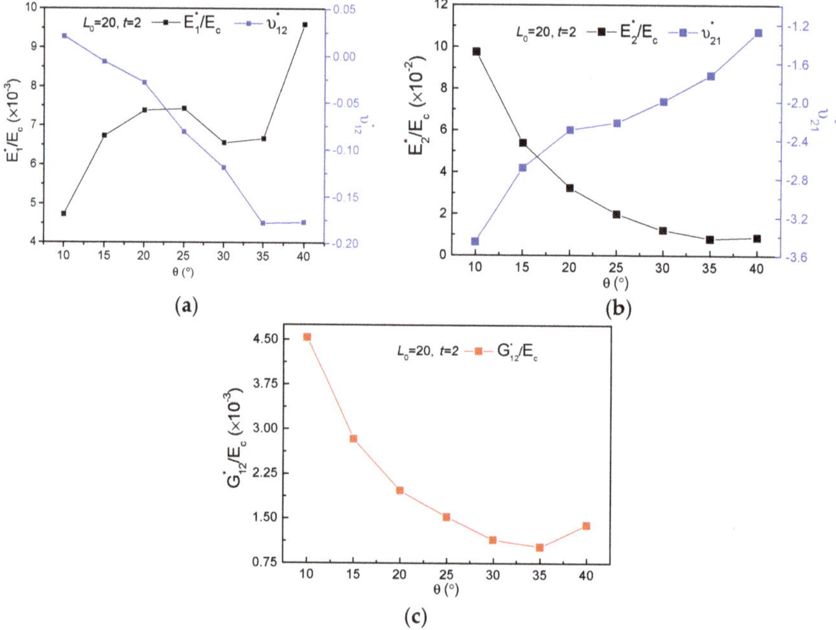

Figure 6. The variations of effective mechanical properties verse inclined angle θ: (**a**) E_1^*/E_c and v_{12}^*, (**b**) E_2^*/E_c and v_{21}^*, (**c**) G_{12}^*/E_c.

In view of increasing cell thickness t mainly enhancing the structural weight and stiffening mechanical behaviors, therefore, the effects of cell inclined angle θ for different L_0 on in-plane mechanical property were studied in detail and are demonstrated in Figure 7. It may be observed that variation and magnitude of the effective mechanical properties for $L_0 = 20$ with enhancive θ are completely different from those of $L_0 = 40 - 120$. With the increasing θ, non-dimensional modulus E_1^*/E_c displays an increasing and decreasing trend for $L_0 = 40, 60$, and exhibits a gradual decrease for $L_0 = 80 - 120$, presented in Figure 7a; E_2^*/E_c and G_{12}^*/E_c are both observed declines with slower slopes, for $L_0 = 40$ to $L_0 = 120$, shown in Figure 7b,c, respectively. For the study of Poisson's ratio, the increasing cell angle θ makes v_{12}^* decrease and v_{21}^* increase inversely, seen in Figure 7d,e. Additionally, parameter L_0 makes no apparent effect on Poisson's ratio, except the scenarios of $L_0 = 20$, where v_{12}^* remains constant in the range of $\theta = 35°$ to $\theta = 40°$ as well as v_{21}^* behaviors a relatively great variation from -3.456 with $\theta = 10°$ to -1.271 with $\theta = 40°$.

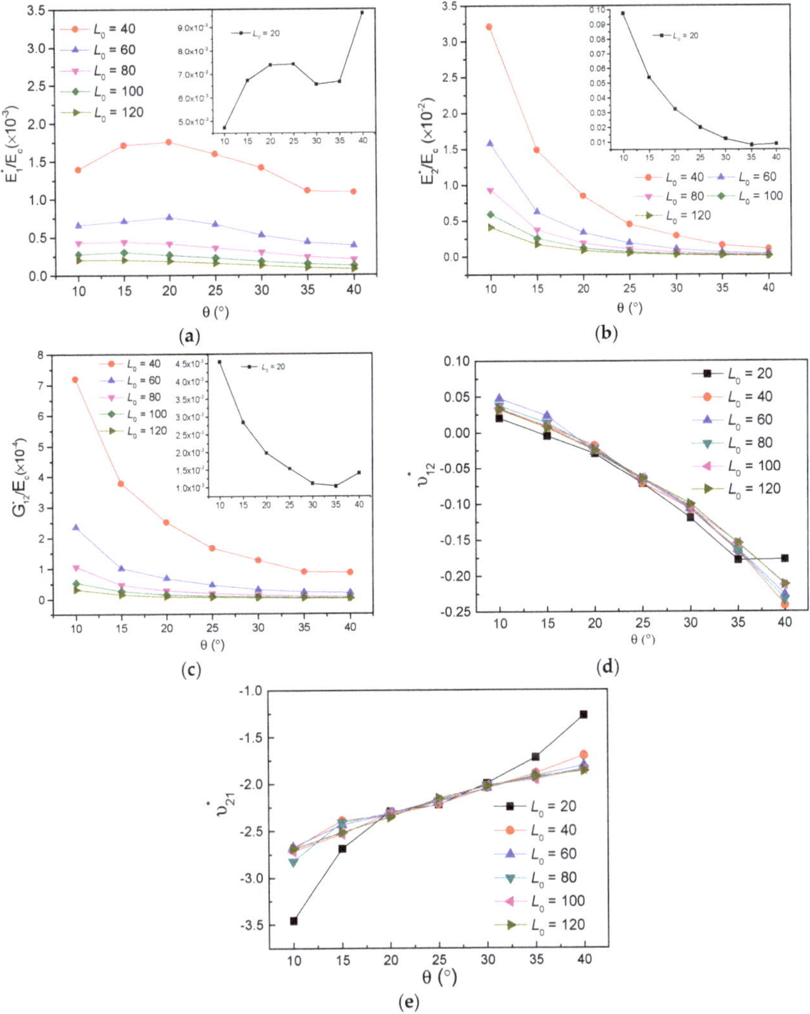

Figure 7. Effects of parameters L_0 and θ on the effective mechanical property for $t = 2$: (**a**) E_1^*/E_c, (**b**) E_2^*/E_c, (**c**) G_{12}^*/E_c, (**d**) v_{12}^*, (**e**) v_{21}^*.

3.2. The Effects of Subordinate Cell Amount

The amount of star subordinate cells along x and y directions are defined and presented in Figure 8, where N_x is the number of half star subordinate cell along x direction and N_y is the number of entire star subordinate cell along y direction. The effects of subordinate cell amount on effective mechanical behavior were then studied. It may be clearly observed from Figure 9a that a growing amount of N_x increases E_1^*/E_c and v_{12}^*, when $N_x \geq 7$, v_{12}^* turns positive. However, increasing N_x makes both E_2^*/E_c and G_{12}^*/E_c decline and remains v_{21}^* to be constant, presented in Figure 9b,c, respectively. Figure 10 shows how the effective properties of the new metamaterials vary with subordinate cell amount along y direction. Non-dimensional effective modulus E_1^*/E_c and G_{12}^*/E_c decrease with the increasing number of N_y, Poisson's ratio v_{12}^* increases and stays auxetic, whereas, E_2^*/E_c and v_{21}^* exhibit the opposite variations under an increasing number of N_y. Consequently, it is found that a small amount of N_x and appropriate number of N_y can satisfy the new hierarchical metamaterial with wholly auxetic behavior and strong stiffness in-plane simultaneously.

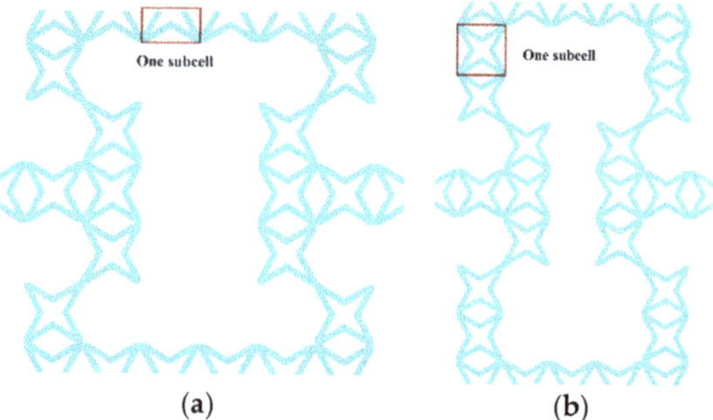

Figure 8. The number variations of star subordinate cell: (**a**) N_x along x direction, (**b**) N_y along y direction.

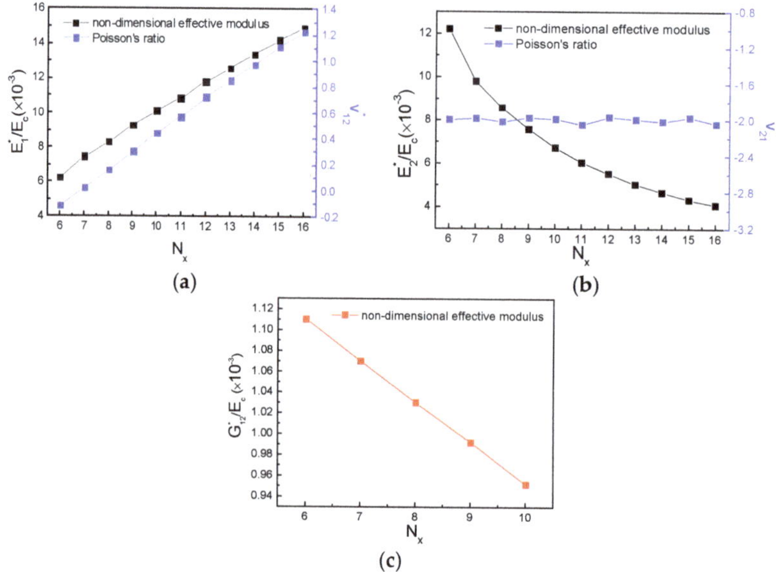

Figure 9. Variation of subordinate cell amount N_x verse: (**a**) E_1^*/E_c and Poisson's ratio v_{12}^*, (**b**) E_2^*/E_c and Poisson's ratio v_{21}^*, (**c**) G_{12}^*/E_c.

3.3. Relative Density Study

The formula of the relative density for one representative unit cell is given by Equation (2); the variation of relative density ρ/ρ_c with parameters L_0, θ and t were obtained and are shown in Figure 11a–c. For Figure 11d, parameter θ was valid in a range of $2.5° - 42.5°$, corresponding ρ/ρ_c exhibits a non-monotonic going up and down variation. Therefore, the values of θ in the range of 35° to 40° made a different influence on the results of relative density as well as previous effective mechanical properties shown in Figure 7.

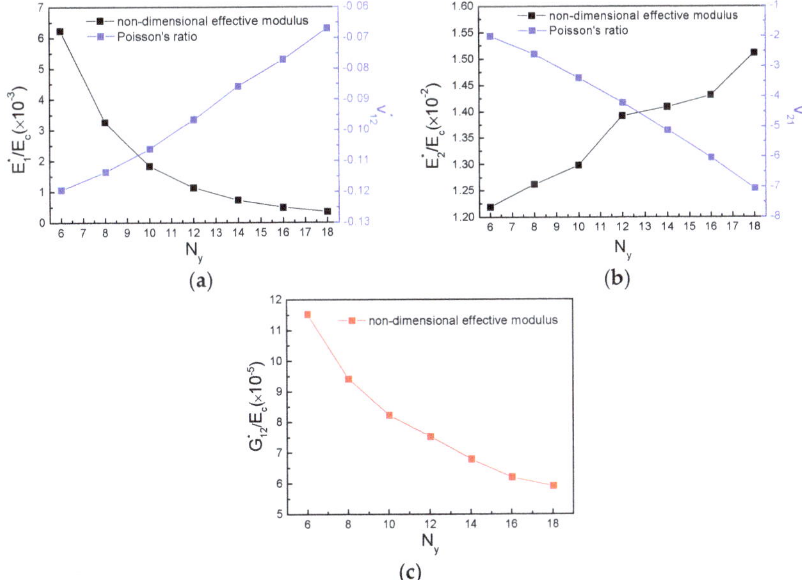

Figure 10. Variation of subordinate cell amount N_y verse: (**a**) E_1^*/E_c and Poisson's ratio v_{12}^*, (**b**) E_2^*/E_c and Poisson's ratio v_{21}^*, (**c**) G_{12}^*/E_c.

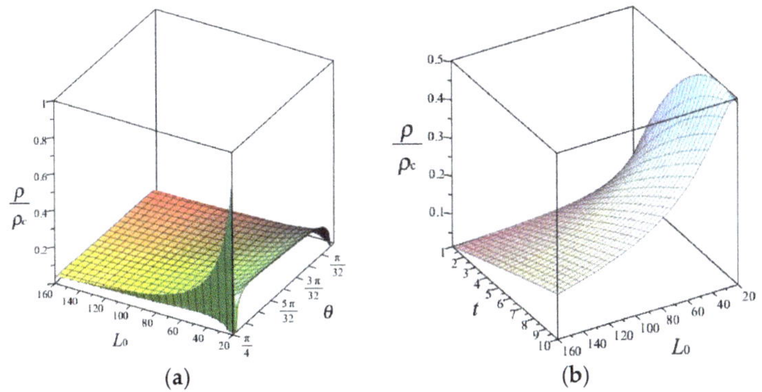

Figure 11. *Cont.*

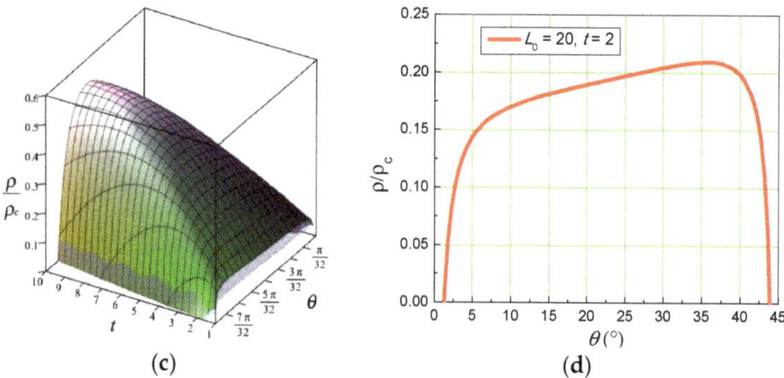

(c)

(d)

Figure 11. Geometric parameters effects on relative density of the hierarchical RUC: (**a**) $t = 2$, (**b**) $\theta = 30°$, (**c**) $L_0 = 20$, (**d**) $L_0 = 20$ and $t = 2$.

The relationship between effective mechanical properties and relative density ρ/ρ_c simulated by FEM finite element method were investigated as following. Variations of specific stiffness $(E_i^*/E_c)/(\rho/\rho_c)$ with different parameters are represented in Figures 12 and 13, respectively. In Figure 12a, specific stiffness all decline with increasing L_0, extremum values of $(E_2^*/E_c)/(\rho/\rho_c) = 0.05963$ and $(G_{12}^*/E_c)/(\rho/\rho_c) = 0.00563$ were achieved with $L_0 = 20$, $\theta = 30°$ and $t = 2$. Similarly, for the increase of cell thickness t, specific stiffnesses all exhibit a growing variation with different gradient, seen in Figure 12b. The impact of parameter θ on effective mechanical behavior of the proposed new metamaterial were investigated and are represented in Figure 12c. For the increase of θ, it may be observed that $(E_2^*/E_c)/(\rho/\rho_c)$ and $(G_{12}^*/E_c)/(\rho/\rho_c)$ have a resembling variation of first decline and then ascent, while $(E_1^*/E_c)/(\rho/\rho_c)$ exhibit an up-down-up variation. In the range of $\theta = 35°$ to $\theta = 40°$, specific stiffnesses $(E_i^*/E_c)/(\rho/\rho_c)$ all increase monotonically.

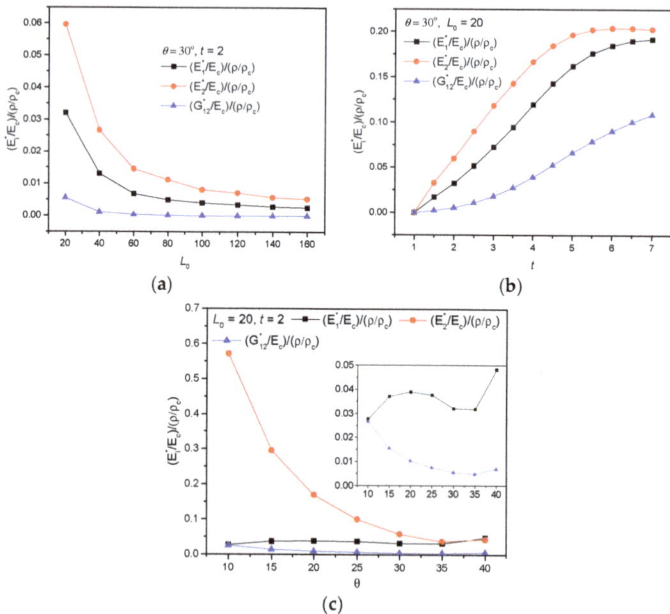

(a)

(b)

(c)

Figure 12. Variation of specific stiffness $(E_i^*/E_c)/(\rho/\rho_c)$ with parameters: (**a**) L_0, (**b**) t, (**c**) θ.

Figure 13 presents the variation of in-plane specific stiffness with the cell angle θ for various parameter L_0, while $t = 2$. The specific stiffness along 1-direction exhibits three different variational trends with a rising θ for different L_0, respectively, shown in Figure 13a: (1) for $L_0 = 20$, $(E_1^*/E_c)/(\rho/\rho_c)$ varies as an up-down-up curve and reaches the maximum value of 0.04829 with $\theta = 40°$, which is 25 times greater than the one for $L_0 = 120$; (2) for $L_0 = 40, 60$, $(E_1^*/E_c)/(\rho/\rho_c)$ presents a first increasing and then decreasing variation; (3) for $L_0 \geq 80$, $(E_1^*/E_c)/(\rho/\rho_c)$ decreases monotonically. From Figure 13b,c, it can be observed that $(F_2^*/F_c)/(\rho/\rho_c)$ and $(G_{12}^*/F_c)/(\rho/\rho_c)$ decrease clearly and then ascend with the variation of cell angle θ from 35° to 40° when $L_0 = 20$, a comparison of the specific stiffness in this range shows that $(E_2^*/E_c)/(\rho/\rho_c)$ varies slightly from 0.03817 to 0.04422, however, $(G_{12}^*/E_c)/(\rho/\rho_c)$ increases significantly from 0.0049 to 0.00704. When $L_0 \geq 40$, $(E_2^*/E_c)/(\rho/\rho_c)$ and $(G_{12}^*/E_c)/(\rho/\rho_c)$ both monotonically decline with increasing θ and L_0. Figure 13c illustrates that an increase of more than 143% of the specific shear stiffness show up when the parameter L_0 varies from 20 to 120 with $\theta = 40°$, which makes varying L_0 also a good design method for $(G_{12}^*/E_c)/(\rho/\rho_c)$. Therefore, it can be determined that all the high specific stiffness in plane can be achieved simultaneously by choosing proper parameter θ and L_0.

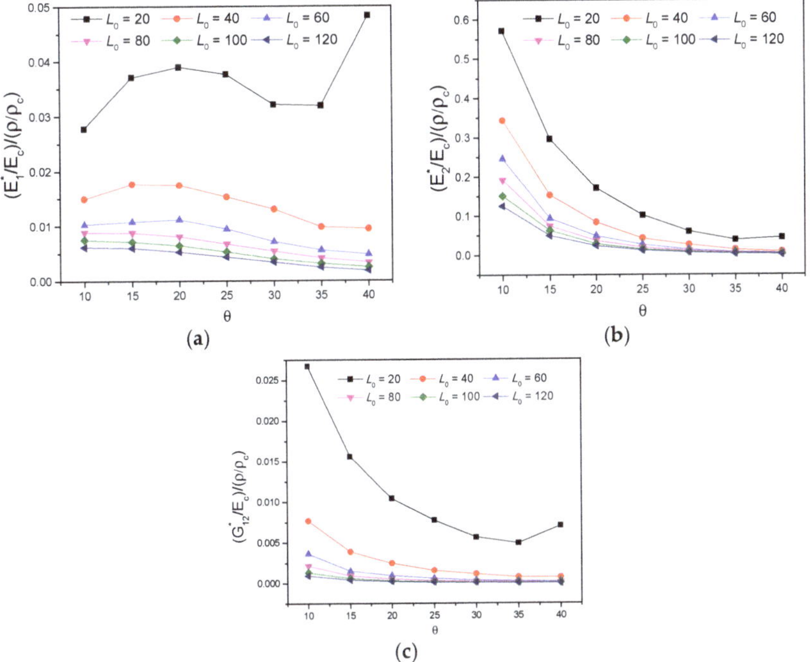

Figure 13. Variations of specific stiffness $(E_i^*/E_c)/(\rho/\rho_c)$ verse parameters L_0 and θ, when $t = 2$: (**a**) $(E_1^*/E_c)/(\rho/\rho_c)$, (**b**) $(E_2^*/E_c)/(\rho/\rho_c)$, (**c**) $(G_{12}^*/E_c)/(\rho/\rho_c)$.

Finally, the ratio between effective Poisson's ratio and relative density verse parameters L_0 and θ were investigated. In Figure 14a, it is seen that $v_{12}^*/(\rho/\rho_c)$ performs from positive to remarkable auxetic behavior and it declines significantly with both increasing L_0 and θ, the maximal descending slope is achieved with $L_0 = 120$. Comparing with Figure 13a, the optimum values of both high specific stiffness and large auxetic deformation in 1-direction can be selected widely for the special curve shape of $(E_1^*/E_c)/(\rho/\rho_c)$ with $L_0 = 20$. Figure 14b reveals that the auxetic $v_{21}^*/(\rho/\rho_c)$ increases with greater θ and decreases with increasing L_0, oppositely. Contrast to Figure 13b, the optimal values of both high specific stiffness and large auxetic deformation in 2-direction is acquired for $\theta = 10°$ and $L_0 = 20$, where $(E_2^*/E_c)/(\rho/\rho_c) = 0.57228$ and $v_{21}^* = -3.45649$.

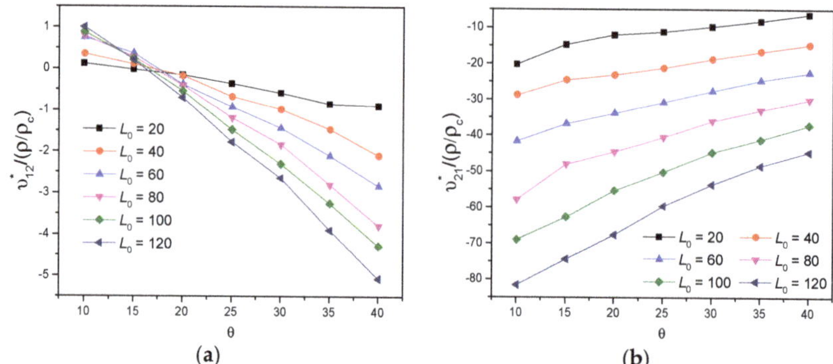

Figure 14. Variations of the ratio between effective Poisson's ratio and relative density verse parameters L_0 and θ, when $t = 2$: (a) $v_{12}^*/(\rho/\rho_c)$, (b) $v_{21}^*/(\rho/\rho_c)$.

4. Conclusions

A novel hierarchical metamaterial with tailorable mechanical properties was proposed using re-entrant planar lattice structure with star-shaped subordinate cell. The effective non-dimensional moduli and Poisson's ratio in plane were simulated by FE homogenization firstly, then the influences of the geometric parameters on mechanical behavior and relative density were studied in detail. It was found that the new hierarchical metamaterial can obtain large variations and control of the design of the in-plane mechanics through the variations of parameters for both the re-entrant RUC and star subordinate cell. Comparing with conventional re-entrant honeycomb, the novel star-re-entrant metamaterial has enhanced mechanical properties of specific stiffness and auxeticity accounting for its hierarchical porosity as well as multilevel tunable parameters. In addition, the new auxetic metamaterial is more convenient fabricated by 3D printing technique as less stress concertation occurs in the connecting tips of star subordinate cell when compared with other zero Poisson's ratio star cellular structure. However, the inclined angle of parent re-entrant RUC is restricted to 45° due to the symmetric simplification of star subordinate cell; as a result, in-plane effective mechanical behavior of the new hierarchical metamaterial can be limited by lacking another internal inclined angle of sub cell. In general, optimum results such as small values of parameters L_0, t and θ for sub cell as well as small amount of N_x and moderate number of N_y for parent RUC can provide the new hierarchical metamaterial with whole auxetic behavior and strong specific stiffness in-plane simultaneously. It is believed that the innovative hierarchical metamaterials will greatly expand the potential applications in the construction, manufacturing and transportation industries due to the inherent low-weight associated with hierarchical systems, like doubly curved panels in aerospace or marine structures. It can also be used in conformable and stretchable electronics, biomedical devices such as porous smart bandage releasing different classes of medications to different extents, as well as the design of smart auxetic stents, etc.

Author Contributions: W.Z., J.W. and F.S. conceived, designed and directed the project. W.Z., S.Z. and R.S. performed the simulations. W.Z. and S.Z. wrote the paper. All authors analyzed the data, discussed the results, and commented on the manuscript.

Funding: This work was funded by Pre-research Foundation of Shenyang Aircraft Design and Research Institute, Aviation Industry Corporation of China (Grant No. JH20128255) as well as Pre-research Key Laboratory Foundation of General Armament Department of China (Grant No. JZ20180035).

Acknowledgments: Wenjiao Zhang would like to thank British Council and Chinese Scholarship Council (CSC) for the funding of her research work through University of Bristol.

Conflicts of Interest: The authors declare no conflicts of interest.

References

1. Gibson, L.J.; Ashby, M.F. *Cellular Solids: Structure and Properties*, 2nd ed.; Cambridge University Press: Cambridge, UK, 1997.
2. Ashby, M.F. The properties of foams and lattices. *Philos. Trans. R. Soc. Lond. Ser. A Math. Phys. Eng. Sci.* **2006**, *364*, 15–30. [CrossRef] [PubMed]
3. Harkati, E.; Daoudi, N.; Bezazi, A.; Haddad, F. Scarpa. In-plane elasticity of a multi re-entrant auxetic honeycomb. *Compos. Struct.* **2017**, *180*, 130 139. [CrossRef]
4. Lira, C.; Innocenti, P.; Scarpa, F. Transverse elastic shear of auxetic multi re-entrant honeycombs. *Compos. Struct.* **2009**, *90*, 314–322. [CrossRef]
5. Scarpa, F.; Tomlinson, G. Theoretical characteristics of the vibration of sandwich plates with in-plane negative Poisson's ratio values. *J. Sound Vib.* **2000**, *230*, 45–67. [CrossRef]
6. Liu, W.; Wang, N.; Luo, T.; Lin, Z. In-plane dynamic crushing of re-entrant auxetic cellular structure. *Mater. Des.* **2016**, *100*, 84–91. [CrossRef]
7. Innocenti, P.; Scarpa, F. Thermal conductivity properties and heat transfer analysis of multi-re-entrant auxetic honeycomb structures. *J. Compos. Mater.* **2009**, *43*, 2419–2439. [CrossRef]
8. Lakes, R. Materials with structural hierarchy. *Nature* **1983**, *361*, 511–515. [CrossRef]
9. Oftadeh, R.; Haghpanah, B.; Papadopoulos, J.; Hamouda, A.M.S.; Nayeb-Hashemi, H.; Vaziri, A. Mechanics of anisotropic hierarchical honeycombs. *Int. J. Mech. Sci.* **2014**, *81*, 126–136. [CrossRef]
10. Mousanezhad, D.; Ebrahimi, H.; Haghpanah, B.; Ghosh, R.; Ajdari, A.; Hamouda, A.M.S.; Vaziri, A. Spiderweb honeycombs. *Int. J. Solids Struct.* **2015**, *66*, 218–227. [CrossRef]
11. D'Alessandro, L.; Zega, V.; Ardito, R.; Corigliano, A. 3D auxetic single material periodic structure with ultra-wide tunable bandgap. *Sci. Rep.* **2018**, *8*, 2262. [CrossRef] [PubMed]
12. Bruggi, M.; Corigliano, A. Optimal 2D auxetic micro-structures with band gap. *Meccanica* **2019**. [CrossRef]
13. Wang, Q.M.; Jackson, J.A.; Ge, Q.; Hopkins, J.B.; Spadaccini, C.M.; Fang, N.X. Lightweight mechanical metamaterials with tunable negative thermal expansion. *Phys. Rev. Lett.* **2016**, *117*, 175901. [CrossRef] [PubMed]
14. Billon, K.; Zampetakis, I.; Scarpa, F.; Ouisse, M.; Hetherington, A. Mechanics and band gaps in hierarchical auxetic rectangular perforated composite metamaterials. *Compos. Struct.* **2017**, *160*, 1042–1050. [CrossRef]
15. Qiao, J.; Chen, C. In-plane crushing of a hierarchical honeycomb. *Int. J. Solids Struct.* **2016**, *85*, 57–66. [CrossRef]
16. Sun, Y.; Chen, Q.; Pugno, N. Elastic and transport properties of the tailorable multifunctional hierarchical honeycombs. *Compos. Struct.* **2014**, *107*, 698–710. [CrossRef]
17. Sun, Y.; Wang, B.; Pugno, N.; Wang, B.; Ding, Q. In-plane stiffness of the anisotropic multifunctional hierarchical honeycombs. *Compos. Struct.* **2015**, *131*, 616–624. [CrossRef]
18. Taylor, C.M.; Smith, C.W.; Miller, W.; Evans, K.E. The effects of hierarchy on the in-plane elastic properties of honeycombs. *Int. J. Solids Struct.* **2011**, *48*, 1330–1339. [CrossRef]
19. Mousanezhad, D.; Haghpanah, B.; Ghosh, R.; Hamouda, A.M.; Nayeb-Hashemi, H.; Vaziri, A. Elastic properties of chiral, anti-chiral, and hierarchical honeycombs: A simple energy-based approach. *Theor. Appl. Mech. Lett.* **2016**, *6*, 81–96. [CrossRef]
20. Gatt, R.; Mizzi, L.; Azzopardi, J.I.; Azzopardi, K.M.; Attard, D.; Casha, A.; Briffa, J.; Grima, J.N. Hierarchical Auxetic Mechanical Metamaterials. *Sci. Rep.* **2015**, *5*, 8395. [CrossRef]
21. Chen, Y.; Jia, Z.; Wang, L. Hierarchical honeycomb lattice metamaterials with improved thermal resistance and mechanical properties. *Compos. Struct.* **2016**, *152*, 395–402. [CrossRef]
22. Yin, H.; Huang, X.; Scarpa, F.; Wen, G.; Chen, Y.; Zhang, C. In-plane crashworthiness of bio-inspired hierarchical honeycombs. *Compos. Struct.* **2018**, *192*, 516–527. [CrossRef]
23. Wu, W.; Tao, Y.; Xia, Y.; Chen, J.; Lei, H.; Sun, L.; Fang, D. Mechanical properties of hierarchical anti-tetrachiral metamaterials. *Extre. Mech. Lett.* **2017**, *16*, 18–32. [CrossRef]
24. Wu, W.; Song, X.; Liang, J.; Xia, R.; Qian, G.; Fang, D. Mechanical properties of anti-tetrachiral auxetic stents. *Compos. Struct.* **2018**, *185*, 381–392. [CrossRef]
25. Gong, X.; Huang, J.; Scarpa, F.; Liu, Y.; Leng, J. Zero Poisson's ratio cellular structure for two-dimensional morphing applications. *Compos. Struct.* **2015**, *134*, 384–392. [CrossRef]

26. Lira, C.; Scarpa, F.; Olszewska, M.; Celuch, M. The SILICOMB cellular structure: Mechanical and dielectric properties. *Phys. Status Solidi* **2009**, *246*, 2055–2062. [CrossRef]
27. Salit, V.; Weller, T. On the feasibility of introducing auxetic behavior into thin-walled structures. *Acta Mater.* **2009**, *57*, 125–135. [CrossRef]
28. Zhang, W.; Neville, R.; Zhang, D.; Scarpa, F.; Wang, L.; Lakes, R. The two-dimensional elasticity of a chiral hinge lattice metamaterial. *Int. J. Solids Struct.* **2018**, *141*, 254–263. [CrossRef]

Article

Effects of Intrinsic Properties on Fracture Nucleation and Propagation in Swelling Hydrogels

Jingqian Ding [1], Ernst W. Remij [1], Joris J. C. Remmers [1] and Jacques M. Huyghe [2,*]

[1] Department of Mechanical Engineering, Eindhoven University of Technology, P.O. BOX 513,
   5600 MB Eindhoven, The Netherlands; jingqian.ding@outlook.com (J.D.); ernst_remij@hotmail.com (E.W.R.);
   j.j.c.remmers@tue.nl (J.J.C.R.)
[2] Bernal Institute, University of Limerick, Limerick, V94 T9PX, Ireland
* Correspondence: Jacques.huyghe@ul.ie

Received: 28 April 2019; Accepted: 23 May 2019; Published: 27 May 2019

Abstract: In numerous industrial applications, the microstructure of materials is critical for performance. However, finite element models tend to average out the microstructure. Hence, finite element simulations are often unsuitable for optimisation of the microstructure. The present paper presents a modelling technique that addresses this limitation for superabsorbent polymers with a partially cross-linked surface layer. These are widely used in the industry for a variety of functions. Different designs of the cross-linked layer have different material properties, influencing the performance of the hydrogel. In this work, the effects of intrinsic properties on the fracture nucleation and propagation in cross-linked hydrogels are studied. The numerical implementation for crack propagation and nucleation is based on the framework of the extended finite element method and the enhanced local pressure model to capture the pressure difference and fluid flow between the crack and the hydrogel, and coupled with the cohesive method to achieve crack propagation without re-meshing. Two groups of numerical examples are given: (1) effects on crack propagation, and (2) effects on crack nucleation. Within each example, we studied the effects of the stiffness (shear modulus) and ultimate strength of the material separately. Simulations demonstrate that the crack behaviour is influenced by the intrinsic properties of the hydrogel, which gives numerical support for the structural design of the cross-linked hydrogel.

Keywords: hydrogel; swelling; fracture; nucleation; propagation

---

1. Introduction

As industrial applications require computational tools that resolve the microstructure of products rather than tools that smooth out microstructures, increasingly higher levels of versatility and robustness are demanded of finite element codes. Indeed, heterogeneities, discontinuities, microcracking, and contact problems can potentially complicate a microstructural computation. As an example of a such a computation, in this paper we study the swelling of hydrogel beads. Swelling or drying volume transitions of ionized hydrogels can be induced by a continuous change of various conditions, such as temperature, pH, electric field, and salt concentration [1]. The degree of volume transformation depends on the composition and structure of the cross-linked network [2,3]. Because of their swelling behaviour, ionized hydrogels have received considerable attention for pharmaceutical and industrial applications, such as drug delivery or disposable diapers. Many swelling processes of hydrogels start from a dry state. When the dry hydrogel is placed in a solvent, it absorbs fluid and swells. During this volume transformation, a stress field is created within the hydrogel, and to some degree, cracks are generated and developed [4,5].

As the US patent US7517586B2 addressed, hydrogel-forming polymers used as absorbents require adequately high sorption capacity and gel strength [6]. Gel strength resists deformation of hydrogel particles during swelling and avoids gel blocking within a swollen gel bed, which is achieved by increasing the level of surface cross-linking. Although a surface cross-linked hydrogel increases gel stiffness at the surface, it reduces the absorbent capacity and increases the tendency to have a brittle fracture. It is crucial to optimize the balance between absorbent capacity and gel strength.

The effect of the microstructure of the material on swelling capacity has been well-tested; however, few studies take the effects of damage into consideration. Cervera et al. [7] developed a computational model to analyse the progressive cracking due to the swelling of concrete in large concrete dams. Zhang et al. [8] numerically studied the phenomenon of crack healing induced by swelling in gels. Guo et al. [9] found out that the intrinsic properties of the interface between a polymeric hydrogel and a rigid substrate have a great influence on the opening profile of the interface crack. Although the studies mentioned above deal with crack behaviours induced by swelling, there is still a long way to go. Considering that the structural optimization of the hydrogel is largely affected by the intrinsic properties of the cross-linked and original hydrogel, it is important to study the effects of material properties on crack behaviours.

The extended finite element method (XFEM) is an efficient tool used to simulate fracture growth without re-meshing. It was first applied by Belytschko and Black [10] by adding an additional degree of freedom on nodes which belong to the element crossed by the discontinuity. Wells and Sluys [11] incorporated a cohesive surface formulation into the method to achieve the crack propagation in any arbitrary direction. Similarly, Leonhart and Meschke took the moisture transport in opening discontinuities into account and analysed the crack propagation in partially saturated porous media [12]. Kraaijeveld et al. [13] took osmotic forces into consideration and studied Mode I crack propagation in saturated ionized porous media in small deformations. Irzal et al. [14] extended the partition of unity approach of fracturing porous media into the finite deformation regime. In order to capture the pressure gradient across the discontinuity, Remij et al. [15] developed the enhanced local pressure model (ELP): a separate degree of freedom for the pressure in the discontinuity was added to the pressure left and right of the crack. Furthermore, the level set method (LSM) [16] is commonly incorporated with XFEM to model crack growth. LSM is used to locate the crack and its tip, and it simplifies the selection of the enhanced nodes in XFEM.

In the present work, a finite deformation model is presented to study the crack behaviour of a heterogeneous swelling hydrogel. We integrated XFEM and ELP to capture the pressure difference and fluid flow between the crack and the hydrogel, and used the cohesive zone method to achieve crack propagation without re-meshing [17].

2. Results

2.1. Crack Propagation

Here, we consider a circular sample consisting of three different materials (Figure 1). We fixed the middle point and constrained the node located on the boundary between the core and middle part at $0°$ in the y direction. A changing chemical potential was applied at the outer surface, which led to the swelling of the medium. The material properties are listed in Table 1.

Figures 2 and 3 show the behaviour of crack propagation with different stiffnesses and ultimate strength of the shell. The crack propagates over time for four different shell shear moduli (2 MPa, 3 MPa, 3.5 MPa and 4.5 MPa) and the same shell ultimate strength of 0.52 MPa (Figure 2). For $G = 2.0$ MPa, the initial crack opens without propagation. For the other four examples, every initial crack propagates. Generally, a stiffer material reaches high stress faster than a softer material with the same amount of deformation, causing material failure and crack growth. The higher the shear modulus is, the earlier the initial crack propagates, as seen in Figure 2 with crack propagation plotting for $G = 3.0, 3.5,$ and 4.5 MPa. At the same time, a stiffer shell resists the particle's deformation and helps to keep the shape of the particle. Less deformation comes with smaller stress in the middle part, which suppresses the crack growth (Figure 4). This is the reason why the crack length of $G = 4.5$ MPa is much smaller than with $G = 3.0$ MPa and $G = 3.5$ MPa.

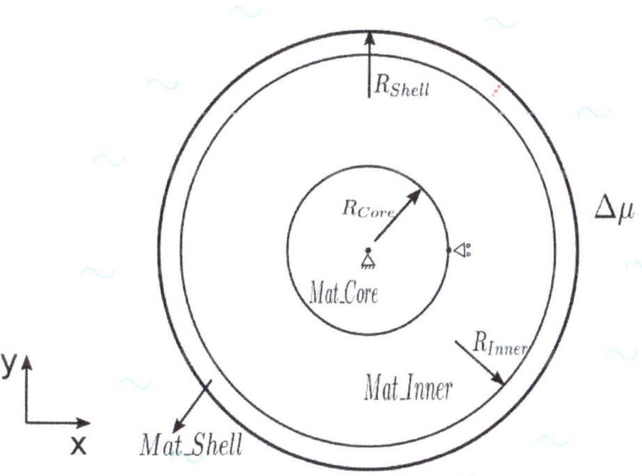

Figure 1. The geometry and boundary conditions of the sample. A changing chemical potential is applied along the outer surface. The red dashed line indicates the initial crack. ($R_{Shell} = 0.5$ mm, $R_{Inner} = 0.45$ mm, $R_{Core} = 0.2$ mm)

Table 1. Material properties.

Name	Symbol	Value
Shear modulus (core)	G	0.05 MPa
Shear modulus (middle)	G	$(1.0, 1.5, 1.7)$ MPa
Shear modulus (shell)	G	$(2.0, 3.0, 3.5, 4.0, 4.5)$ MPa
Intrinsic permeability	k_{int}	1.0×10^{-11} mm^2
Fluid dynamic viscosity	μ	1.0×10^{-9} MPa s
Porosity	φ	0.83
Ultimate strength (core)	τ_{ult}	0.01 MPa
Ultimate strength (middle)	τ_{ult}	$(0.3, 0.4, 0.5, 0.6)$ MPa
Ultimate strength (shell)	τ_{ult}	$(0.48, 0.52, 0.54)$ MPa
Toughness	G_c	0.01 N/mm
Gas constant	R	8.3145 J$\cdot$mol$^{-1}\cdot$K^{-1}
Temperature	T	293.0 K
Initial fixed charge density	c_0^{fc}	332.0×10^{-6} mmoleq/mm^3
External salt concentration	c_{ex}	154.0×10^{-6} mmol/mm^3

Similarly, we fixed the shell shear modulus ($G_{shell} = 4$ MPa), and studied the effect of various ultimate strengths (0.3 MPa, 0.4 MPa, 0.5 MPa, and 0.6 MPa) of the middle material (Figure 3). The effect of ultimate strength is straightforward; the ultimate strength does not contribute to the propagating length of the crack, but it increases the capacity of the material to resist tension, and only affects the rate of propagation. The higher the ultimate strength, the later the crack starts to propagate. It is advisable to achieve relatively high ultimate strength in the material design to obtain higher elongating resistance.

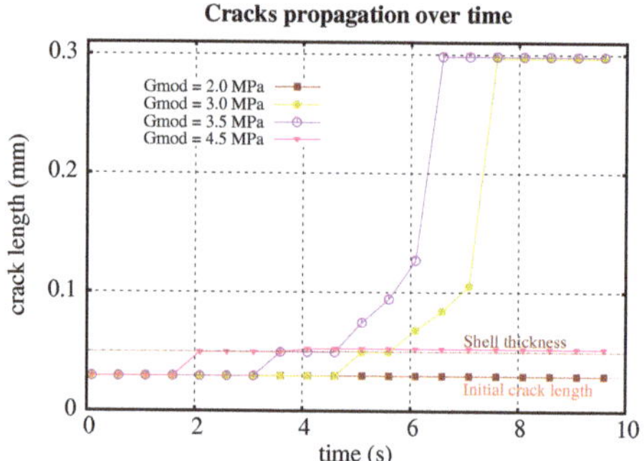

Figure 2. Crack propagation profile over time with different stiffnesses of the shell.

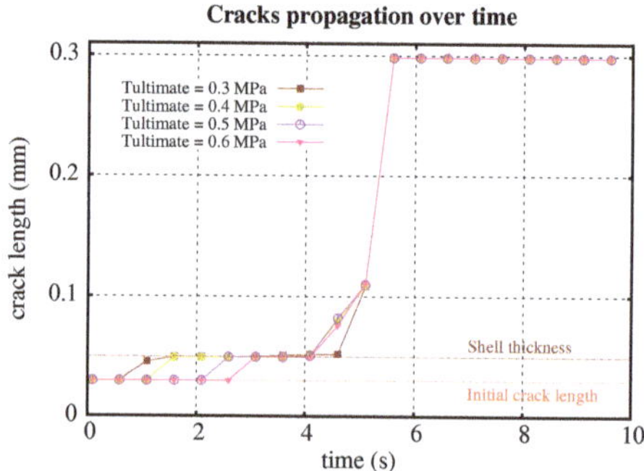

Figure 3. Crack propagation profile over time with different ultimate strengths of the shell.

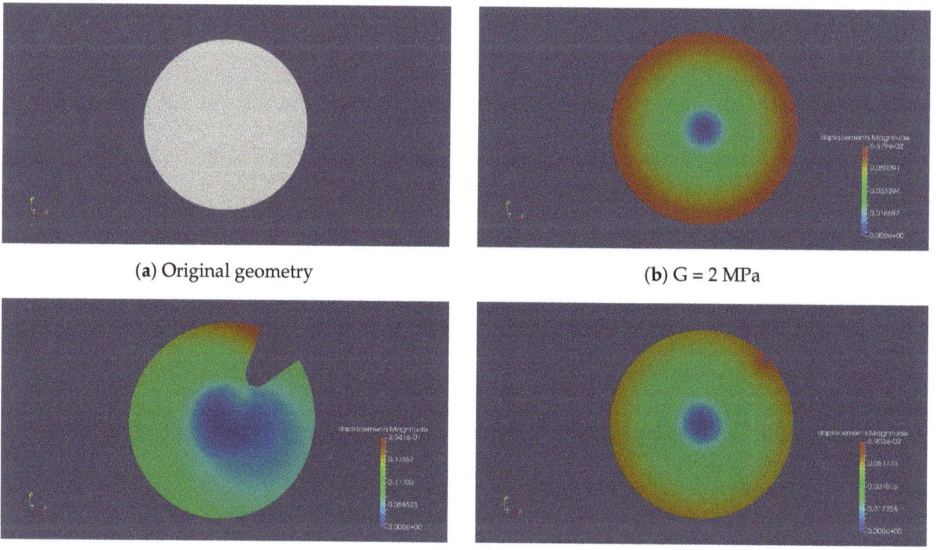

(**a**) Original geometry

(**b**) G = 2 MPa

(**c**) G = 3 MPa

(**d**) G = 4.5 MPa

Figure 4. The displacement profile with different shear moduli of the shell.

2.2. Crack Nucleation

The energy accumulates within the hydrogel particle when it swells. There are two ways to dissipate energy, propagate existing cracks, or nucleate new cracks. In Section 2.1 we discussed the behaviour of crack propagation. In the current section, we discuss the behaviour of crack nucleation.

The same geometry (Figure 1) and material properties (Table 1) are used here to model crack nucleation. In order to avoid too many cracks nucleating at the same time too close together, we made an extra constraint that there were at least 30 elements between two cracks. We compared the nucleation state within three groups: different shear moduli of the shell, different ultimate strengths of the shell, and different shear moduli of the middle part of the sample.

The point at $45°$ to the x-axis in the first quadrant is the initial crack. In Figure 5, we plotted the nucleations with the shell's ultimate strength of 0.48 MPa, 0.52 MPa, and 0.54 MPa, respectively. There was no nucleation with an ultimate strength of 0.54 MPa. For $\tau_{ult} = 0.52$ MPa, it has five nucleations. When τ_{ult} decreases to 0.48 MPa, the nucleations increase to 7.

Figure 6 shows the nucleation locations with different shear moduli of the shell. There were nine nucleations for $G = 4.5$ MPa and five nucleations for $G = 4.0$ MPa. No new cracks nucleated for $G = 3.0$ MPa. Similarly, Figure 7 plots the nucleation locations with different shear moduli of the middle part of the particle. It shows that there are 7, 5, and 0 nucleations relating to $G = 1.0, 1.5$ and 1.7 MPa, respectively. The distribution of nucleations is roughly symmetric about the diameter crossing the initial crack.

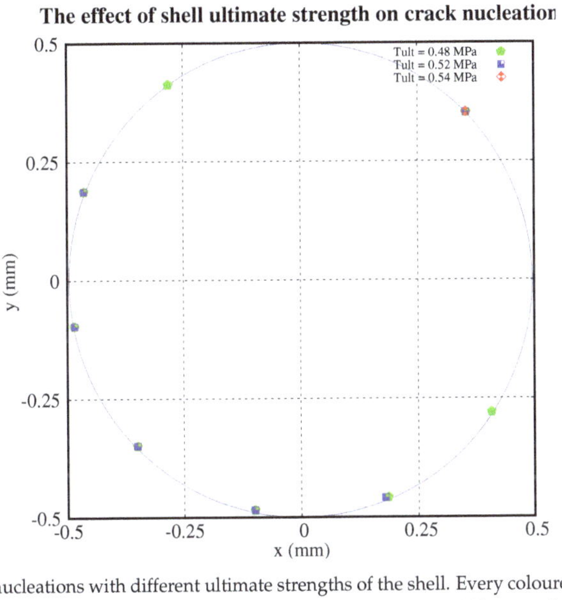

Figure 5. Crack nucleations with different ultimate strengths of the shell. Every coloured dot represents one nucleation site. The results of three separate computations with different values of the shell's ultimate strength are superimposed. The colour of the dot specifies the computation to which it belongs.

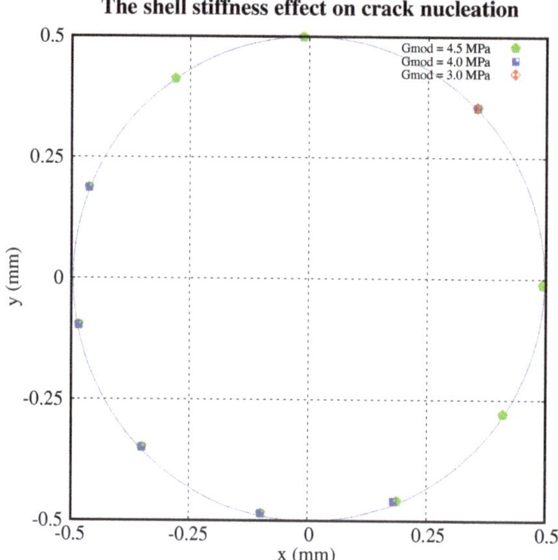

Figure 6. Crack nucleations with different stiffnesses of the shell.

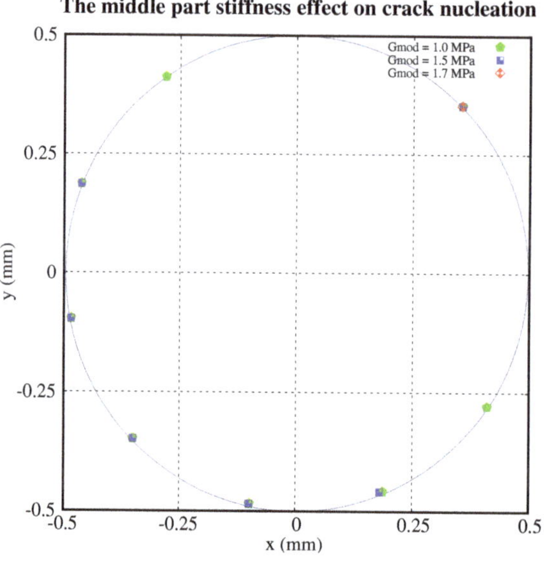

Figure 7. Crack nucleations with different stiffnesses of the middle part.

Figure 8 is the chemical potential distribution within the crack with different time steps. The nucleation process shows a cascade phenomenon (Figure 8). At the time of 4.8 s, the initial crack propagates without any nucleation. After 0.1 s, there is one nucleation. Another new crack nucleated after 0.1 s. At the time of 5.1 s, the previous two nucleations kept growing, and another three new cracks were generated.

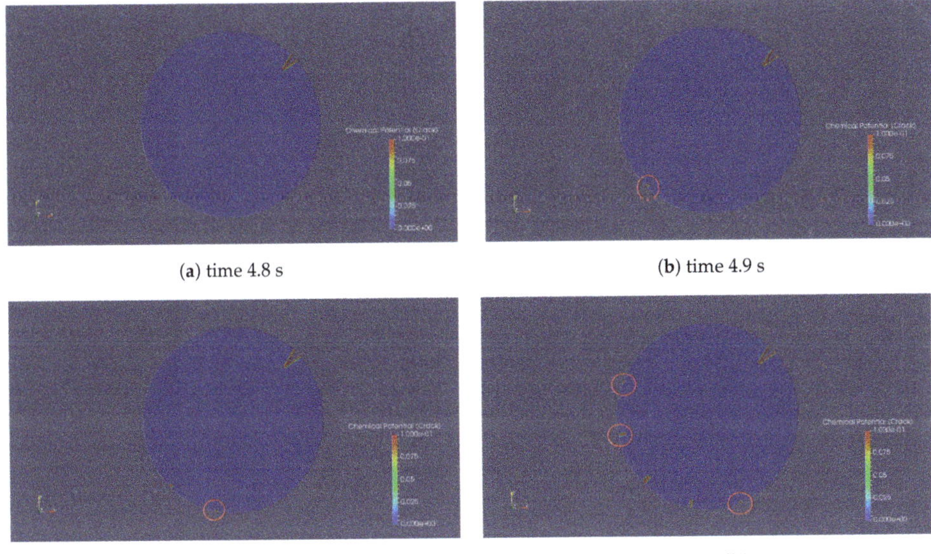

(**a**) time 4.8 s   (**b**) time 4.9 s

(**c**) time 5.0 s   (**d**) time 5.1 s

Figure 8. Nucleations at different times (s), red circles emphasize nucleation spots.

3. Discussion

This study illustrates how the relationship between the microstructure and function of a product can be studied using a dedicated nonlinear multiphysics, multicomponent, finite element analysis. The type of analysis done in this study was made possible with commercial codes, because they lack the vital options needed. The feasibility of complex non-linear analyses combining large deformations and mesh-free multiple crack propagations through a swelling heterogeneous material has been demonstrated. Much more is needed for the robustness of the XFEM code in computations that resolve the microstructure, as deformations are very large, locally, and heterogeneities hamper the smoothness of the solutions. This investigation is mainly focused on the effects of intrinsic properties on the fracture behaviour. We studied two dissipative mechanisms here, where one is the propagation of existing cracks, and the other is the nucleation of cracks. From the results, it appears that the crack propagation and nucleation are largely affected by the intrinsic properties of the material, and particularly the properties of the cross-linked shell around the softer swelling hydrogel particle.

Generally, the higher the shear modulus of the shell, the earlier the initial crack propagates. Besides, a stiffer shell resists the particle's ability to deform, and helps to keep the shape of the particle. Less deformation comes with smaller effective stress in the middle part, which suppresses crack growth. However, a swelling media requires high swelling capacity. Therefore, there needs to be a balance between the swelling of the inner part and the elastic stiffness of the outer part. This numerical simulation can be used as a tool to optimize the material property of the gel in the microstructural design of the swelling hydrogel particle. By comparing the crack propagation with different ultimate strengths of the shell, we found that the ultimate strength only affects the rate of the propagation. The higher the ultimate strength is, the slower the crack propagates.

From Figures 5–7, we conclude that:

- The higher the ultimate strength of the shell, the fewer the cracks which nucleate.
- The higher the shear modulus of the shell, the more cracks which nucleate.
- The higher the shear modulus of the middle part of the particle, the fewer the cracks which nucleate.
- The distribution of nucleations is roughly symmetric about the diameter crossing the initial crack.

Figure 8 presents a cascade of nucleations with different crack openings with material properties of Table 1. The events illustrate that the failure of the gel builds up in stages. It starts from fewer defects and weakens the material while the material is still functioning. When more and more defects appear and interact with each other, the material finally fails. The process of the defects is the same as the process of the nucleation. The opening of cracks not only depends on the stress state, but also on the neighbouring cracks. If the neighbouring crack is located close to the current crack with a relatively large opening, it will impede the opening of the current crack.

The software presented in this paper serves as a great numerical support for the design of a proper cross-linked shell, such as the special cross-link density required to achieve a specific stiffness and fracture resistance. The stiffness ratio between the inner part and the shell is a critical parameter that determines the performance of the product. A high ratio delays the swelling of the hydrogel, and too high a ratio disallows the failure of the hydrogel altogether.

4. Methods

4.1. Kinematic Relations

We considered a body Ω crossed by a discontinuity (Figure 9). The body was divided into two subdomains, Ω^+ and Ω^-. The total displacement field of the solid skeleton was described by a regular displacement field $\hat{\mathbf{u}}$ and an enhanced displacement field $\tilde{\mathbf{u}}$,

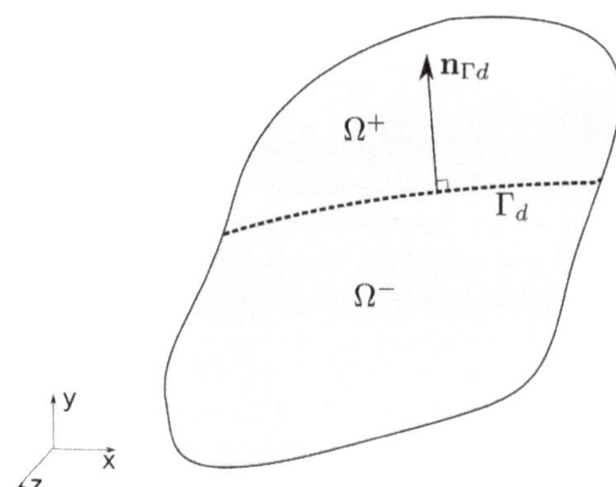

Figure 9. The body Ω is crossed by a discontinuity (dashed line). $\mathbf{n}_{\Gamma_d}$ represents the normal of the discontinuity surface pointing to Ω^+.

$$\mathbf{u}(\mathbf{X}, t) = \hat{\mathbf{u}}(\mathbf{X}, t) + \mathcal{H}_{\Gamma_d}(\mathbf{X})\tilde{\mathbf{u}}(\mathbf{X}, t), \tag{1}$$

where $\mathbf{X}$ is the material point in the reference configuration of the solid, $\mathcal{H}_{\Gamma_d}$ is the Heaviside step function, defined as

$$\mathcal{H}_{\Gamma_d} = \begin{cases} 1 & \mathbf{X} \in \Omega^+ \\ 0 & \mathbf{X} \in \Omega^- \end{cases}. \tag{2}$$

The chemical potential field is discontinuous across the discontinuity and the hydrogel, and defined as

$$\mu^f(\mathbf{X}, t) = \hat{\mu}^f(\mathbf{X}, t) + \mathcal{H}_{\Gamma_d}(\mathbf{X})\tilde{\mu}^f(\mathbf{X}, t), \tag{3}$$

In the discontinuity, the chemical potential is equal to an independent variable μ_d,

$$\mu^f = \mu_d, \quad \mathbf{X} \in \Gamma_d. \tag{4}$$

Hence, the value of the chemical potential jumps from $\hat{\mu}^f$ to μ_d to $\hat{\mu}^f + \tilde{\mu}^f$ as one crosses the discontinuity from Ω^- to Ω^+.

4.2. Balance Equations

We considered the body as a solid skeleton with fully saturated interstitial fluid. It was assumed that there was no mass transfer, and thermal gradients, inertia, and gravity were neglected. Based on Biot's theory, the momentum balance reads

$$\nabla \cdot \sigma = 0 \quad \text{in} \quad \Omega, \tag{5}$$

with σ the total stress, which is decomposed into the effective stress σ_e and the pore fluid pressure p,

$$\sigma = \sigma_e - p\mathbf{I}, \tag{6}$$

with $\mathbf{I}$ being the unit tensor.

Equations (5) and (6) can be written with respect to the reference configuration, using the transformation of $\mathbf{P} = J\sigma \cdot \mathbf{F}^{-T}$, read

$$\nabla_0 \cdot \mathbf{P} = 0 \quad \text{in} \quad \Omega_0 \quad \text{(momentum balance)},$$
$$\mathbf{P} = \mathbf{P}_e - Jp\mathbf{F}^{-T} \quad \text{(total first Piola-Kirchhoff stress)},$$

where $\mathbf{P}_e$ is the effective first Piola-Kirchhoff stress.

Conservation of mass for an incompressible fluid yields the mass balance in the reference configuration,

$$\dot{J} + \nabla_0 \cdot \mathbf{Q} = 0, \tag{7}$$

with $\dot{J} = J\text{div}\dot{\mathbf{u}}$ and $\mathbf{Q} = -\mathbf{K} \cdot \nabla_0 \mu^f$ the seepage flux obeying the Darcy's relation in the presence of the concentration gradient. In the equation of the seepage flux, $\mathbf{K}$ is the permeability tensor back-transformed to the reference configuration, and μ^f is the chemical potential, defined as

$$\mu^f = p - \pi, \tag{8}$$

with p being the hydrostatic pressure, and π the osmotic potential.

The fracture process behaviour is governed by a traction separation law. Here, we assumed stress continuity from the gel to the discontinuity, the local momentum balance being described as

$$\mathbf{P} \cdot \mathbf{n}_{\Gamma_d} = J||\mathbf{F}^{-T} \cdot \mathbf{n}_{\Gamma_d}||\mathbf{t}_d - J(\mu_d^f + \pi_d)\mathbf{F}^{-T} \cdot \mathbf{n}_{\Gamma_d}, \tag{9}$$

in which $\mathbf{n}_{\Gamma_d}$ is the normal of the discontinuity Γ_d, $\mathbf{t}_d$ is the traction, and μ_d^f and π_d are the chemical potential and osmotic pressure, respectively, within the discontinuity.

The local mass balance was obtained by integrating the continuous mass balance across the discontinuity.

$$\mathbf{n}_{\Gamma_d} \cdot (\mathbf{Q}_{\Gamma_d}^+ - \mathbf{Q}_{\Gamma_d}^-) = Ju_n \frac{\partial}{\partial s}(k_d \frac{\partial \mu_d^f}{\partial s}) - Ju_n \langle \frac{\partial v}{\partial s} \rangle - J\dot{u}_n, \tag{10}$$

where k_d is the conductivity in the discontinuity

$$k_d = \frac{(u_n)^3}{12\mu}, \tag{11}$$

with μ being the viscosity of the fluid, $u_n = \tilde{\mathbf{u}} \cdot \mathbf{n}_{\Gamma_d}$ the opening of the discontinuity.

4.3. Swelling Behaviours

Swelling equilibria between the hydrogel and the immersed fluid must fulfil $\mu_{in} = \mu_{ex}$ (μ_{in} is the chemical potential within the hydrogel, and μ_{ex} describes the chemical potential of the surrounding fluid).

The difference of the osmotic pressure between the hydrogel and the surrounding fluid allows ions or fluid to go into or out of the hydrogel, leading to swelling or shrinking. Because the ionic diffusion coefficient of SAP is two orders of magnitude larger than the pressure diffusion coefficient of SAP, we assumed the ionic constituent to be drained. Based on Van's Hoff empirical relation, the osmotic pressure difference was given by

$$\Delta\pi(\varepsilon) = RT\sqrt{(c^{fc})^2 + 4(c^{ex})^2} - 2RTc^{ex},\tag{12}$$

where R is the gas constant, T is the temperature, c^{fc} is the fixed charge density, and c^{ex} is the external salt concentration.

4.4. Nucleation Mechanism

Crack nucleation and growth are part of the fracture development. Nucleated micro-cracks are based on the stress state of the solid skeleton. The stress state was obtained by averaging effective stresses around the crack tip. Remmers et al. [18] used a Gaussian weighted function to calculate the averaged stress:

$$\sigma_{av} = \sum_{i=1}^{n_{int}} \frac{\omega_i}{\omega_{tot}}\sigma_i$$

$$\omega_{tot} = \sum_{j=1}^{n_{int}} \omega_j,$$

here, n_{int} is the number of integration points in the domain, and ω_i is the weight factor relating to the integration point i. The weight factor is defined as

$$\omega_i = \frac{1}{l_a^3}e^{\frac{-r_i^2}{2l_a^2}},\tag{13}$$

where l_a is the length scale parameter, and r_i denotes the distance between the integration point i and the crack tip.

4.5. Crack Propagation Mechanism

The crack propagation is governed by the stress state of the crack tip. The stress is calculated by averaging the stress around the crack tip. The traction $\mathbf{t}_d$ in Equation (9) acting on the fracture surface is based on the cohesive constitutive relation, and governs the propagation of the crack. When the averaged stress of the crack tip exceeds the ultimate strength, the crack starts to propagate. The normal traction t_n is described as

$$t_n = \tau_{ult}exp(-\frac{u_n\tau_{ult}}{\mathcal{G}_c}),\tag{14}$$

where τ_{ult} is the ultimate strength of the material, and $\mathcal{G}_c$ denotes the fracture toughness.

4.6. Constitutive Equation

In this paper, we used a compressible Neo-Hookean model. Though the solid itself was incompressible, due to its porous structure, the entire solid matrix was compressible. The Cauchy stress [19,20] is given by

$$\sigma = -\frac{1}{6}\frac{\ln J}{J}G\mathbf{I}\left(-1 + \frac{3(J + n_{s,0})}{(-J + n_{s,0})} + \frac{3\ln(J)Jn_{s,0}}{(-J + n_{s,0})^2}\right) + \frac{G}{J}(\mathbf{F} \cdot \mathbf{F}^T - J^{2/3}\mathbf{I}), \tag{15}$$

where G is the shear modulus, and $n_{s,0}$ represents the initial solid volume fraction. As J tends to $n_{s,0}$, the fluid content vanishes and Equation (15) becomes an incompressible law.

When osmotic swelling is included, the total stress reads

$$\sigma_{tot} = \sigma - (\mu^f + \pi)\mathbf{I}. \tag{16}$$

Author Contributions: conceptualization, J.M.H.; methodology, J.D. and E.W.R.; software, J.D. and E.W.R.; validation, J.D.; formal analysis, J.D.; investigation, J.D.; resources, J.J.C.R.; data curation, J.D. and J.J.C.R.; writing—original draft preparation, J.D.; writing—review and editing, J.M.H. and J.D.; supervision, J.M.H. and J.J.C.R.; project administration, the Dutch Technology Foundation STW and J.M.H.; funding acquisition, the Dutch Technology Foundation STW, the technological branch of the Netherlands Organization of Scientific Research NWO and the Ministry of Economic Affairs.

Funding: This research was funded by the Dutch Technology Foundation STW, the technological branch of the Netherlands Organization of Scientific Research NWO and the Ministry of Economic Affairs under contract [12538] Interfacial effects in ionized media.

Acknowledgments: We acknowledge the financial support from the Dutch Technology Foundation STW, the technological branch of the Netherlands Organization of Scientific Research NWO and the Ministry of Economic Affairs under contract [12538] Interfacial effects in ionized media.

References

1. Tanaka, T. Phase Transitions of Gels. In *Polyelectrolyte Gels*; American Chemical Society: Washington, WA, USA, 1992; Chapter 1, pp. 1–21.
2. Ikeda-Fukazawa, T.; Ikeda, N.; Tabata, M.; Hattori, M.; Aizawa, M.; Yunoki, S.; Sekine, Y. Effects of crosslinker density on the polymer network structure in poly- N, N -dimethylacrylamide hydrogels. *J. Polym. Sci. Pol. Phys.* **2013**, *51*, 1017–1027. [CrossRef]
3. Saunders, J.R.; Moussa, W. Dynamic mechanical properties and swelling of UV-photopolymerized anionic hydrogels. *J. Polym. Sci. Pol. Phys.* **2012**, *50*, 1198–1208. [CrossRef]
4. Baumberger, T.; Caroli, C.; Martina, D. Solvent control of crack dynamics in a reversible hydrogel. *Nat. Mater.* **2006**, *5*, 552–555. [CrossRef]
5. Tanaka, Y.; Shimazaki, R.; Yano, S.; Yoshida, G.; Yamaguchi, T. Solvent effects on the fracture of chemically crosslinked gels. *Soft Matter* **2016**, *12*, 8135–8142. [CrossRef]
6. Fossum, R.D.; Schmidt, M.; Meyer, A. Absorbent Structures Comprising Coated Water-Swellable Material. U.S. Patent 7517586B2, 14 April 2009.
7. Cervera, M.; Oliver, J.; Herrero, E.; Oñate, E. A computational model for progressive cracking in large dams due to the swelling of concrete. *Eng. Fract. Mech.* **1990**, *35*, 573–585. [CrossRef]
8. Zhang, J.; An, Y.; Yazzie, K.; Chawla, N.; Jiang, H. Finite element simulation of swelling-induced crack healing in gels. *Soft Matter* **2012**, *8*, 8107. [CrossRef]
9. Guo, J.; Luo, J.; Xiao, Z. On the opening profile and near tip fields of an interface crack between a polymeric hydrogel and a rigid substrate. *Eng. Fract. Mech.* **2016**, *153*, 91–102. [CrossRef]
10. Belytschko, T.; Black, T. Elastic crack growth in finite elements with minimal remeshing. *Int. J. Numer. Methods Eng.* **1999**, *45*, 601–620. [CrossRef]
11. Wells, G.; Sluys, L. A new method for modelling cohesive cracks using finite elements. *Int. J. Numer. Meth. Eng.* **2001**, *50*, 2667–2682. [CrossRef]
12. Leonhart, D.; Meschke, G. Extended Finite Element Method for hygro-mechanical analysis of crack propagation in porous materials. *Proc. Appl. Math. Mech.* **2011**, *11*, 161–162. [CrossRef]

13. Kraaijeveld, F.; Huyghe, J.M.; Remmers, J.J.C.; de Borst, R. Two-Dimensional Mode I Crack Propagation in Saturated Ionized Porous Media Using Partition of Unity Finite Elements. *J. Appl. Mech.* **2013**, *80*, 020907. [CrossRef]

14. Irzal, F.; Remmers, J.J.; Huyghe, J.M.; De Borst, R. A large deformation formulation for fluid flow in a progressively fracturing porous material. *Comput. Methods Appl. Mech. Eng.* **2013**, *256*, 29–37. [CrossRef]

15. Remij, E.W.; Remmers, J.J.C.; Huyghe, J.M.; Smeulders, D.M.J. The enhanced local pressure model for the accurate analysis of fluid pressure driven fracture in porous materials. *Comput. Methods Appl. Mech. Eng.* **2015**, *286*, 293–312. [CrossRef]

16. Stolarska, M.; Chopp, D.L.; Mos, N.; Belytschko, T. Modelling crack growth by level sets in the extended finite element method. *Int. J. Numer. Meth. Eng.* **2001**, *51*, 943–960. [CrossRef]

17. Ding, J.; Remmers, J.; Leszczynski, S.; Huyghe, J. Swelling driven crack propagation in large deformation in ionized hydrogel. *J. Appl. Mech.* **2018**, *85*, 021007. [CrossRef]

18. Remmers, J. Discontinuities in Materials and Structures. Ph.D. Thesis, Delft University of Technology, Delft, The Netherlands, 2006.

19. Wilson, W.; Huyghe, J.M.; van Donkelaar, C.C. A composition-based cartilage model for the assessment of compositional changes during cartilage damage and adaptation. *Osteoarthr. Cartil.* **2006**, *14*, 554–560. [CrossRef] [PubMed]

20. Wilson, W.; Huyghe, J.M.; Van Donkelaar, C.C. Depth-dependent compressive equilibrium properties of articular cartilage explained by its composition. *Biomech. Model. Mech.* **2007**, *6*, 43–53. [CrossRef] [PubMed]

MDPI

St. Alban-Anlage 66

4052 Basel

Switzerland

Tel. +41 61 683 77 34

Fax +41 61 302 89 18

www.mdpi.com

Polymers Editorial Office

E-mail: polymers@mdpi.com

www.mdpi.com/journal/polymers